SOLVING THE QUANTUM PUZZLE
PARADIGM CHANGE IN MILLISECONDS
VOLUME 1 – UNDERSTANDING THE COGNITIVE PROBLEM

"Therefore, the seeker after the truth is not one who studies the writings of the ancients and, following his natural disposition, puts his trust in them, but rather the one who suspects his faith in them and questions what he gathers from them, the one who submits to argument and demonstration, and not to the sayings of a human being whose nature is fraught with all kinds of imperfection and deficiency. The duty of the man who investigates the writings of scientists, if learning the truth is his goal, is to make himself an enemy of all that he reads, and attack it from every side. He should also suspect himself as he performs his critical examination of it, so that he may avoid falling into either prejudice or leniency."

Abu Ali ibn al-Haytham aka 'Al-Hazen'
about the scientific method (around the year 1025)

This book is dedicated to my son Eric, who followed this project with interest for many years (and suddenly was a head taller than me), and my old friend Abed in Copenhagen. In 2005, he provided me with his comfortable studio and a 'desk' in his café, which is known for its first-class Mediterranean and international cuisine, during the preparation of the manuscript. If you should ever come to Copenhagen, drink a really good coffee or eat well, just ask for Café Phenix in Valby. Without his selfless support and almost self-evident tolerance of an 'artistic passion' that devours so much time without the slightest chance of success that most people would hardly understand, this book would probably never have seen the light of day. (This book was first published in German in 2008 under the title 'Quantum Top Secret').

Note in 2023: The book presented here is a new edition, revised, expanded and translated into English, consisting of two volumes. Volume 1 "Understanding The Cognitive Problem" will be published in November 2023. Volume 2, "The Discovery of Molecular Cell Division and the Branching of Light", describes the development of 20th century quantum physics, they key experiments, its interpretive problems and derives from them a "New Copenhagen Interpretation". To be published in spring 2024.

MARIO WINGERT

SOLVING THE QUANTUM PUZZLE

PARADIGM CHANGE IN MILLISECONDS

VOLUME 1

UNDERSTANDING THE COGNITIVE PROBLEM

Bibliographic information of the German National Library: The German National Library lists this publication in the German National Bibliography; detailed bibliographic data are available on the Internet at dnb.dnb.de.

© Mario Wingert 2023
1th edition November 2023
272 pages with 34 illustrations and figures
contact: avoga@proton.me
Production and publishing: BoD – Books on Demand, Norderstedt
ISBN: 9783757809201

Contents

Preface: Farewell to the atom
A long overdue paradigm change

If one day we really understand quantum physics, it will be even more
revolutionary than the achievements of Copernicus and Columbus –
and for everyone, not just us physicists. (Anton Zeilinger, 2005)

Why physicists and chemists should read this book as well

Twenty years ago, in 2003, I published a small booklet in German
with the title "Einsteins Vermächtnis: Die Revolution der Physik –
Die Lösung des Wellen-Teilchen-Paradoxons"[1]. Although this title
should arouse the curiosity of any full- blooded physicist, I hardly
expected a reaction, because it was clear to me that most physicists
have not believed for many decades that the quantum puzzle is still
solvable. Since then, most physicists have more or less come to terms
with the so-called Copenhagen Interpretation of NIELS BOHR, who
in 1927 had had to conclude that atoms, elementary particles and
waves cannot really exist at the deepest level of nature. At the same
time, however, he had also claimed that it was impossible on principle to design fundamentally new models of the true constitution
of nature. This enigmatic interpretation of the cognitive problems
of physics has puzzled professional physicists and laymen alike, and
has spawned mystical speculations about the nature of reality and
the role of the 'conscious observer' that are actually unworthy of a
scientific approach.

Today, almost a hundred years later, one can only state that physics has still not succeeded in solving its conceptual problems, which
prevent a genuine understanding of the physical nature of light and
matter. With the body idea of mechanics, the way we understand
motion in mechanics, and the atomic hypothesis, we are unable to
understand the true constitution of matter, which turns out to be
divisible in a mysterious, seemingly incomprehensible way. In a similar but opposite way, the wave model fails when one seeks to understand the same division processes with single light rays or individual
emitted and absorbed light quanta, electrons, atoms and molecules.

---

1 Translated: "Einstein's Legacy: The Revolution of Physics – Solving the Wave-Particle-Paradox"

However, Quantum Mechanics, the branch of physics which deals with these problems, succeeds in circumscribing these phenomena in a mathematically suitable way by alternately applying the particle and the wave model, although they actually contradict each other. So physics has managed to approach reality, even if the true face of nature still remains hidden behind a veil. What these equations really mean for our understanding of nature and whether and how our scientific world view must change as a result has remained unclear until today.

The cognitive block triggered by the quantum mechanical interpretation is not entirely innocent of this, for quantum mechanics does not only provide a neutral mathematical description of the phenomena: Since the logical and physical contradictions do not disappear thereby, an accompanying interpretation is still needed, which tries to convince all students of quantum physics that the wave-particle contradiction is not resolvable and deeper physical insights are generally impossible. This was at least the conviction of NIELS BOHR, WERNER HEISENBERG and MAX BORN, the creators of the quantum mechanical formalism, which later became also known as Copenhagen Interpretation. However, a deep uneasiness remains with all physicists, because the sensual contradictions do not disappear. So, in the end, every physicist has to ask himself why he could not solve the tricky quantum puzzle despite his sharp mind.

As a creative designer, I would immediately suspect that most physicists share one or more fundamental assumptions that they believe to be correct and experimentally proven, but which must in fact be false. One of these assumptions, which is clearly refuted by the double-slit experiment, is the atom hypothesis, the assumption of indivisibility. Nevertheless, nobody dares to throw the atomic concept overboard, because it is considered indispensable for science. This is exactly what a paradigm is. Since it is apparently no longer possible for professional physicists to fundamentally question the atom paradigm without running the risk of committing professional suicide, I will assume this role. Paradigmatic thinking, by the way, is a typical product of conformity and group-think and the mortal enemy of creativity, scientific-technical progress and democracy. It was to protect us from such aberrations and matters of faith that the scientific method was originally introduced. Of course, there are always situations where it seems very difficult and sometimes even impossible to find out the cause of apparent or actual contradictions. This leads to phases of stagnation in the progress of science, which can only be overcome by non-obvious, truly creative approaches to solutions. However, it is exactly this possibility that the interpretation of BOHR has categorically excluded, despite of EINSTEIN's warnings. In this book, however, I want to show that BOHR's pessimistic attitude is surmountable if the scientific method is followed consistently and optimistically. This inevitably leads to a new, experimentally well-founded hypothesis about the nature of light and matter, which neither BOHR nor EINSTEIN ever considered: This is the hypothesis of *holistic divisibility*, which follows inevitably from experiment.

In the mid 1920s, physics was still in a revolutionary phase in which creative thinking led to rapid advances in knowledge: Within only four years, between 1923 and 1927, it turned out that the particle conception of mechanics and the atomic hypothesis could not apply to reality. This did not only apply to the atoms of the chemical elements, which in the double-slit experiment would have to be divisible as a whole, which directly contradicted the original meaning of the word 'atom' – *the indivisible*. This was also true for the supposedly indivisible elementary particles, called electrons and protons, of which the atom should consist. In short, the atomic hypothesis, i.e. the assumption that there are indivisible particles in reality, turned out to be wrong – a paradigm that failed already at the deepest level of nature, just when one tried to understand more precisely the constitution of the most elementary mass particle, the hydrogen atom. All these particles turned out to be divisible in a mysterious way, as it was known until then only from light and the resulting interference phenomena. The hypothetical elementary mass particles could therefore not be tiny billiard balls, since they could basically divide and behave like light waves, which led to the fact that one began to apply wave models also to matter.

However, twenty years earlier EINSTEIN had already made it clear that the continuous electromagnetic field theory and the associated wave model of light must be incomplete, since they could not represent the true structure of light and the quantum-like manner in which light energy is transmitted to matter. He suspected already in 1905 that the *quantum of action* postulated by PLANCK mathematically has a real physical meaning and points to a non- continuous energetic structure of light rays and electromagnetic fields. The true nature of this structure was not yet clear to EINSTEIN; so it could be granular or *cellular* (my hint). He explicitly illustrated this structure provisionally with indivisible tiny energy particles, called light quanta, knowing very well that the interference properties of light definitely exclude all particle ideas already since YOUNG's double-slit experiment of 1805. This was exactly the reason why no physicist was ready to accept EINSTEIN's hypothesis of light for twenty years: The contradiction with the experiment was too obvious. EINSTEIN had originally hoped that the wave and interference properties of light could one day be derived from the superposition of separated quantities of energy concentrated in tiny spheres. Such volume elements would have allowed a nice transition to HUYGEN's principle, the kinetic gas theory, and to statistical mechanics. In 1923, with the COMPTON experiment, the hope flickered again briefly that the light quantum could be understood as an energy particle and *atom of light* after all, for which LEWIS coined the term *photon* in 1926.

At the same time, however, it became clear – at least for HEISENBERG and his fellows in Göttingen (Germany), but not for all physicists – that EINSTEIN's provisional atomistic conception of the light quantum met the same fate as the particle idea of the electron and atom: It could not apply to reality, since one could be sure in the meantime that even individual emitted light quanta, electrons and atoms would take two paths at the same time in double-slit experiments, i.e. would have to be divisible. So if such volume elements should really exist, they would have to be divisible at the double-slit and in similar interference experiments, which excludes any atomistic interpretation. After that, however, they would have to reunite and act as a whole at only one point. Since this divisibility was independent of the energy intensity of light beams or the density of the matter beams, theoretically even single rays, waves or wave pulses carrying only the respective minimum amount of energy (i.e. corresponding to individually emitted light quanta, electrons or atoms), should be able to divide and reunite. This appeared extremely puzzling, apparently made no physical sense and contradicted all *known* theories of physics.

EINSTEIN's actual revolutionary hypothesis that light rays and electromagnetic fields in general could have an energetic structure of whatever kind, which was also responsible for the point-like actions of light radiation, however, proved to be extremely successful. Thus only the provisional particle model had to be abandoned, but not the structure hypothesis of light rays and of electromagnetic fields. As BOHR had had to realize in 1913, the concept of light quanta was essential for explaining the structural changes of the hydrogen atom (the more appropriate term is *integral molecule*, as we will see later). The structure of this mass particle changed apparently holistically by the emission and absorption of single light quanta, which BOHR modeled mechanically with orbiting and jumping electrons. With this model he was able to explain the BALMER series of spectral lines emitted by hydrogen. Twelve years later, however, it became finally clear that mechanical models – i.e. models based on the assumption of moving bodies – were no longer tenable. The problem was only that no physicist had an idea how to understand the true structure of light and matter when the indivisibility hypothesis, the particle concept, the motion concept of mechanics and the wave theory fail at the same time.

Nevertheless, it already became clear that light and matter must be physically so similar that they would have to reveal the same constitutional principles in double-slit experiments: On the one hand, a still unknown kind of divisibility, which leads to the well-known interference phenomena, which until then could only be modeled with wave theory. However, this wave theory fails in the description of the absorption of the energy quantity divided at the double slit, because the total emitted energy always comes to effect only at one point *as a whole*, and not at two. From this, one can only conclude that the energetic wholeness of the local effect is logically compatible with the experimentally proven divisibility *only* if one considers the atomic hypothesis as experimentally refuted. In other words:

The double-slit experiment refutes the atomistic world view.

And with it also the naive assumption that we can grasp the nature of reality, i.e. the structural constitution of light and matter, additively with bodies in the sense of mechanics ('billiard balls') or volume elements (spheres) in the sense of Euclidean geometry. That is at least my logical conclusion. Of course, Heisenberg, Bohr and Einstein must have seen the dilemma as well, but they interpreted it quite differently, each in his own way. The real question, however, is whether we accept experiment as the judge of physical theories and find new ideas about the 'physique of nature' necessary.

Obviously, physicists were not able until today to accept this actually obvious logical conclusion without restrictions as physical reality. This is the main cause of all problems of understanding which surround quantum theory since almost one hundred years. The insight that the experiment disproves the atom hypothesis is the key to the solution of the quantum puzzle: If one starts from this fundamental insight, the Gordian knot almost untangles itself. However, this simple logical conclusion still collides with an interpretation of the wave theory which saw the light of day in 1905. In the meantime, it belongs to the most important foundations of contemporary physics and decides *on the admissibility of new physical ideas*. Knowing this, the blockade of cognition and the reluctance to accept new physical ideas becomes at least a little more understandable: I'm talking about Einstein's special theory of relativity. Of all things, this theory prevents us from recognizing and acknowledging a light beam, which branches into two partial rays in the experiment, as a physically coherent field structure – although Einstein had postulated a still unknown field structure in his provisional quantum theory. This is quite ironic, or rather dramatic: Einstein could never discover the structure of radiation he had postulated with his quantum hypothesis of light, because his special theory of relativity prevents exactly that, because it is based on a presupposition which assumes a physical and spatial separation between individual light rays. Such an assumption naturally excludes holistic structural concepts from the outset. However, this presupposition was not alone Einstein's idea: It can be traced back to an indivisibility assumption for single light rays and waves, which is hidden in the foundations of the *transverse* wave theory.

The purpose of the transverse wave theory, introduced two hundred years ago, was to explain the opposite polarization of partial waves or rays that results *from the branching* of light beams on glass bodies. Until then, light was understood as longitudinal waves, i.e. as pulsating light waves and rays that behave in principle like sound and pressure waves. So the original question was how the polarization of these light rays comes about, which could well have had to do with the branching process itself. Since nobody thought in this direction, the transverse wave concept seemed to be the only reasonable explanation for the phenomenon of polarization. In this way, the actually obvious idea that even single light rays or waves with lowest possible intensity must branch and dipolarize at the double-slit was axiomatically excluded in the transversal wave theory. The same should now be true for any kind of radiation modeled by wave theory, including light quanta, electrons and atoms. In this way, the wave theory seemed to reinforce the fundamental reductionist idea of mechanics that 'things' are composite. And with it also the idea of mechanical divisibility, which, however, if only one ray of light was left, also led to the assumption of indivisibility, at least in a *transversal* wave theory. So it seemed that single light rays cannot split. This explains why EINSTEIN found the atomistic particle model of the light quantum so seductive at first, although it contradicted experiment. And with this short reference to the history of ideas in physics, I have already sketched the *second hidden assumption*, which is the reason, besides the atomic hypothesis, why the quantum puzzle seems so tricky and unsolvable. Now we may understand a little better why one automatically ends up in a blind alley if one cannot (or does not want to) critically question the atomic hypothesis, the transverse wave theory and the special theory of relativity.

So, in order to solve the quantum puzzle once and for all, it must also be clarified whether special relativity is still tenable at all if the underlying indivisibility assumption of single light rays or waves is wrong. Such a bold question is (besides the doubt about the physical sense of the Copenhagen interpretation) the second point at which professional physicists would run the risk of committing suicide – or drawing their swords against imaginary enemies of science. Either way, leave the swords in the scabbard and rather think twice:

It is not about disproving EINSTEIN, but to grow beyond him. EIN-STEIN's deep insight that the principle of relative motion must apply both to matter and light is the starting point also for us. So we have to ask ourselves whether the relative principle cannot be interpreted differently, what EINSTEIN's definition of a constant speed of light has to do with the assumption of separability, and whether there are still unchecked assumptions hidden behind the logic of EINSTEIN. For this one must know that special relativity does not question the wave model at all, but only modifies the definition of the speed of light, in such a way that the principle of relative motion is also valid for electromagnetic induction. Note further that there is no connection between EINSTEIN's special theory of relativity and EINSTEIN's quantum hypothesis – although they originated in the same year –, since one is based on a wave model and the other assumes an unclear structure model. These two models have yet to be made compatible. How this can be achieved, and how it affects the interpretation of the principle of relativity, the nature of light and new physical models, reveals the solution of the quantum puzzle: What could not be recognized until today is that the physical divisibility shown in the experiments must be a kind of divisibility still unknown in physics, namely a non-mechanical, *holistic division process*, as we know it so far only from biological structures. This is the unexpected solution of the quantum puzzle that does not fit any expectation grid – a real discovery, at least for me, that now makes possible a physical and vivid understanding of the long sought 'quantum mechanism'. Of course, this is not an ordinary mechanism in the sense of mechanics, but a new physical principle: It characterizes the elementary process of physical structure formation as *branching*, bifurcation, or *cell division*.

I first outlined how this structure formation can be understood in my little booklet of 2003, which had not received much attention. It was all the more surprising for me that this inconspicuous book was recommended as professional literature in the Einstein Year 2005 (in Germany), together with books by JOHN STACHEL and ANTON ZEILINGER. However, I don't know how it came about. I don't know any physicist and dialogues didn't take place either. At least it was a sign of appreciation that encouraged me to continue.

So I wrote a second book, "Quantum Top Secret: Die Lösung des Quantenrätsels. Metamorphose eines Weltbildes" [2], first published in 2008 (in German), in which I explained my ideas and their connection to the history of the ideas in physics and chemistry in a more popular scientific way. In contrast to the general reader, however, most physicists again did not seem particularly interested. This is quite paradoxical: The general reader desperately wants to understand the quantum puzzle and the world view of modern physics, but the quantum mechanic lacks the faith to do so. However, I must also admit that some of my hints about possible *switches of cognition* were at first only intuitive guesses inspired by the experiments.[3] These may have been too vague for theoretical physicists, but their main purpose was to inspire them to conduct their own investigations in the *history of science*. Having been able to review and study more thoroughly a number of original papers in recent years, I am now in a position to name these switches of cognition fairly accurately, with appropriate references and original citations. Besides – I had also to realize that my ideas do not seem as simple and obvious to many educated readers as they do to me. Apparently no one has ever thought so radically about the conceptual problems of physics.

A typical example is AVOGADRO's molecule hypothesis, which owes its origin to chemical reactions of gases, which disproved DALTON's atom hypothesis already in 1811. As far as I know, no chemist, no physicist, and no quantum mechanic ever noticed that this was the first experimentally founded refutation of the indivisibility hypothesis. I believe that if EINSTEIN, BOHR, HEISENBERG, PAULI, HUND and MULLIKEN had been aware of this historical fact, they would have immediately agreed on one thing: With the hypothesis of holistic divisibility, the molecular theory remains valid even if the atomic hypothesis fails. So there is no reason to doubt the existence of an objective reality only because the atomic hypothesis and the body ideas of mechanics (and those of our mind) fail at the constitution of nature and reality. This is just the typical fate of physical hypotheses, models and theories that turn out to be wrong in experiments.

---

2 Translated: Solving the Quantum Puzzle. Metamorphosis of a World View

3 Original phrase in German: "Weichen der Erkenntnis"

Similar conceptual problems have to be solved in Maxwell's field and wave theory and Einstein's theory of special relativity, since they still interpret the propagation of light as unidirectional motion in the sense of mechanics. However, the new physical understanding shows that quantum theory is not about motion in the sense of classical mechanics, at least not primarily. Instead, we are concerned with movements in the sense of structural change, which must be modeled bidirectionally and mirror-symmetrically, since branching processes produce *enantiomorphic forms* (i.e., holistic structures with an intrinsic left-right symmetry). This kind of motion already contains the principle of relativity, explains the true nature of magnetic spin and gives a new physical meaning to the molecular orbital theory of Mulliken and Hund, which since 1932 demands from us to explain the molecular structure of matter *completely without atoms*!

In other words: The mysterious 'eigenvalue quantizations' of the quantum physical wave equations, of which until today no physicist knows what they are supposed to mean physically, describe holistic self-structuring processes of matter and light. They have the same physical nature as cell division and branching processes in biology.

So if we put the pieces of the puzzle together correctly, we can recognize the true face of Nature. The realization that it is the atomic paradigm that has blocked the progress of knowledge in physics for almost a hundred years, but that the molecular hypothesis remains valid, is without doubt a great discovery. Insights that so radically overturn traditional conceptions are, of course, in particular need of explanation. For this reason, I have once again taken the time to thoroughly revise, streamline, and translate my ideas into English. I believe that with the new book now available, anyone – whether a skeptical quantum mechanic or a curious reader – can understand the true adventure of science: It is about real cognition through resolution of contradictions. This also requires descriptive, conceptual, and reflective philosophical thinking. This also makes it clear that no scientist is in possession of absolute truth, no matter how famous he may be. And this book also makes clear why we need a revolution of physics: To ensure that the evolution of physics continues.

Mario Wingert, January 2023

I. The tricky quantum case
Paradigm change in milliseconds

The question we still haven't answered is what quantum theory actually tells us about nature. What is its deeper meaning? The brightest minds have been thinking about it for almost a century, and we still have no answer. For me, this means that there is still far too little radical thinking. We'll probably have to throw beloved concepts overboard first. Which ones, I do not know. But the way forward must be entirely revolutionary. (Anton Zeilinger, 2003)

A great discovery that revolutionizes our world view

This book is about solving the quantum puzzle, also known as the wave-particle or wave-quantum paradox. Contrary to a common myth, you don't have to be a physicist or rocket scientist to understand the quantum problem and its solution. Between you and me, physicists don't understand it either – neither the problem nor the solution. Even mathematical knowledge is not required. The only thing you need is an open mind, imagination and an eye for contradictions. I support you with new pictorial conceptions and illustrations of the key experiments, which are actually quite simple. These experiments will guide us through an great intellectual adventure that resembles a detective story: At first, second and third glance, the tricky quantum case seems to make no sense at all. During the investigation, all suspects seem to be honest and highly respected members of society. All have watertight alibis and best references. Nevertheless, in this case, irreconcilable contradictions suggest that one or more of these alibis must be false and that one or more of the suspects must have committed the crime. In our scientific case, the suspects are deeply rooted assumptions about the constitution of nature. They can boast the best references of honorable scientific societies, have allegedly been thoroughly experimentally tested and thus appear so untouchable that they seem to be true beyond any reasonable doubt. And yet one or more of these assumptions must be false. But which ones? Which basic physical assumptions must be put on trial and thrown overboard? In his Sherlock Holmes detective stories, CONAN DOYLE beautifully sums up how to proceed:

> *"Once you eliminate the impossible, whatever remains, however improbable, must be the truth".*

That is exactly what we're going to do in this book.

What physicists find so improbable, even outrageous, despite all the experimental evidence, is that the atom hypothesis, i.e. the indivisibility assumption, could simply be wrong. But that's not all. Equally unsettling and breathtaking is the fact that quantum experiments prove that there must be holistic interrelationships in the universe which can act independently of time, i.e. instantaneous, over arbitrary large distances. This fact, called *non-locality*, challenges EINSTEIN's special theory of relativity, or more precisely, his definition of the speed of light, which presupposes a spatial-physical separation between two rays of light. Yet both theories have best references: At present, they are considered as foundations of modern physics that nobody dares to doubt. Too great is their authority, and too great is the fear that the shiny skyscraper of physics could collapse.

This fear, however, is unfounded. The principle of relative motion remains indispensable, even if we gain new insights into the nature of light and matter. And the molecule theory survives, even if the atomic hypothesis is wrong. Without the paradigmatic assumption of indivisibility, we can explain the constitution of matter and light much better, in full agreement with experiment and our everyday experience. Although it has long been experimentally proven that even single light quanta (photons), electrons, and atoms must take both paths simultaneously in double-slit and two-way interference experiments, it was not clear until now what significance this fact has for physics, for our ideas of the constitution of nature. My discovery (or at least my new hypothesis) is, that these must be *holistic division and branching processes*, which clearly show that the experiment refutes the atom hypothesis and all particle models.

The nature of holistic division processes is explained by a new physical principle, which for the first time gives a physical foundation to the purely mathematically introduced superposition postulate of quantum mechanics. I called it *enantiomorphic branching* or *bifurcation* in my little book of 2003, but the term *polarization* does as well – if one understands this term physically as mirror-symmetric structure formation. And this new physical principle also explains the true nature of magnetic spin. It states that nature, through holistic processes of division and branching, generates holistic 'fields' or forms which are in-themselves oppositely structured. To understand this sentence, take a look at yourself: Your left and right hand are not identical, they are mirror symmetrical forms of each other. You probably just haven't noticed that your two arms are practically mirror-symmetrical branches with five twigs each. The fact that biological structures arise from cell division and branching processes, in which a whole divides and yet remains a whole, is well known. What is new and breathtaking, however, is that this principle now also applies to so-called elementary particles, atoms and molecules, which leads straight to the discovery of *molecular cell division*. Thus, the new physical principle explains the structure formation of matter and light universally and can be verified in physics, chemistry and biology: All matter structures, whether simple or complex, solid or liquid, gaseous or field-like, animate or inanimate, are originally generated by branching and cell division processes. In this way it becomes clear what this quantum crime case is all about. It is about a paradigm change that is long overdue: Quantum physics requires us to abandon the atomistic- mechanistic world view, revolutionize physics and move to a more natural, structural view of nature.

Of course, such a discovery may be met with stubborn rejection or plain ignorance if physics has not yet properly understood the scope of the quantum problem itself. This is normal, because skepticism is essential for the progress of science (after all, I also use this method): So don't believe anything, check all my claims against the experiments and draw your own informed conclusions. However, there is a psychological obstacle: It becomes difficult, if not impossible, to accept that the atom hypothesis is a physically unfounded paradigm if one considers the terms *atom* and *mass particle* as synonyms and confuses them with each other. Because then one believes that with the disappearance of the atom (hypothesis) also the matter disappears from the world – that the 'objective' reality vanishes into thin air.

I think this is exactly what happens to everyone at the first moment. And this also explains the fruitless discussions between EINSTEIN, BOHR and HEISENBERG about the strange nature of reality, in which they desperately tried to understand the failure of the wave model, the atom hypothesis and the particle idea in quantum physics. But actually it is only a question of accepting the refutation of the indivisibility assumption by the experiment, even if one has still no idea how the nature of matter and light is to be interpreted then. With this insight, first of all, the strategy becomes clear: Obviously, all empirical findings of physics and chemistry must remain valid even if the atom hypothesis fails. This means that our models and theories must be wrong or at least incomplete, even those that are considered highlights of physics and chemistry. Then it only remains to clarify what exactly is wrong with them and which of their underlying assumptions cannot be correct. Although this clarification process would actually be the task of all natural scientists, I will do my best to point out exactly where the *switches of cognition* are hidden in the history of physics and chemistry. Speaking of setting a course: All it takes to change a deeply rooted paradigm is just a click in the mind, as DONELLA MEADOWS so beautifully explains:

"You could say paradigms are harder to change than anything else about a system, and therefore this item should be lowest on the list, not the highest. But there's nothing physical or expensive or even slow about paradigm change. In a single individual it can happen in a millisecond. All it takes is a click in the mind, a new way of seeing. Of course individuals and societies do resist challenges to their paradigm harder than they resist any other kind of change..."[4]

To this I would only add that this resistance in a liberal understanding of science is nothing bad or reactionary, as long as it is not poisoned by matters of faith or financial self-interest that have nothing to do with the matter at hand, namely the pursuit of knowledge and scientific progress. Otherwise, this resistance is nothing more than a natural precaution of the enlightened human community to protect us from false prophets, demagogues and ideologues, thus fulfilling a necessary role in the process of cognition.

4   Donella Meadows: Leverage Points. Places to Intervene. 1997

The quantum puzzle: Waves, particles or a still unknown third?

The quantum puzzle consists of a seemingly irresolvable contradiction between our two fundamental physical models of the constitution of nature: The wave model of light and the body model of mechanics. If the bodies in question are small or tiny, they are called corpuscles or particles. In the 19th century it was thought that matter is made up of indivisible particles called atoms, and that light is made up of electromagnetic waves. Light and matter appeared to be two completely different things. But in the beginning 20th century, after many decades of thorough experimental research, it became again clear that matter and light, despite their different manifestations, must in principle have the same physical nature (this was originally MAXWELL's idea). Matter and light showed a kind of structure, which, however, could no longer be explained with particles. This was extremely puzzling, because with it the atom hypothesis and the body idea of the thinking mind seemed to fail in a strange way at the true constitution of nature. The actual problem seemed so incomprehensible that some leading physicists even believed that the human mind had come up against an insurmountable barriere of cognition.

But the real message of nature was that it operated on fundamentally different physical principles that didn't have much to do with moving particles or bodies. In other words: NEWTON's mechanics failed the moment when one tried to apply it to the constitution of matter (two hundred years earlier it had already failed on the constitution of light). For this one must know that everything what has to do with the movement of bodies is called mechanics. Since NEWTON had thoroughly studied the laws of motion of bodies and had abstracted them into mathematical laws with which the motion of stones, cannonballs and celestial bodies could be described equally exactly, it had long been believed that these laws could be applied universally to nature. Since it was further assumed that the bodies themselves consist of particles, Newtonian mechanics should apply even to the smallest particles, the atoms. But this dream had now collapsed – but why, that could hardly be understood: Why should the proven Newtonian physics fail at the constitution of matter, if the smallest particles of matter, the atoms, are nothing else than tiny bodies? Even though the failure of Newtonian mechanics could not yet be understood, it became clear step by step that there must be a still unknown structural principle responsible for the same physical properties of light and matter. This had to have something to do with the strange division processes in the key experiments that could only be modeled with wave theory, which also suggested that the atomic and mechanistic world view might be wrong. However, hardly anyone wanted to seriously consider this possibility, since the atomic hypothesis was also considered to be secure in physics since 1910 at the latest. In this context, quantum physics was born.

However, the associated quantum theory never managed to really understand the common nature of light and matter and to discover the new physical principle behind it. It was only clear that the two 'classical' models and theories of the constitution of nature, waves and indivisible particles, contradict each other, because waves and light beams are divisible, but atomic particles are not. In the absence of new, problem-solving ideas, quantum mechanics is still forced to use both physical models, albeit alternately and mutually exclusive. This is the unresolved contradiction in our fundamental models of the constitution of nature, the so-called wave-particle paradox. And since light acts always in certain energy amounts, called light quanta, one also speaks of the wave-quantum paradox. Strictly speaking, modern physics is about the natural-philosophical and sensual contradiction between the physical properties *wholeness* and *divisibility*, which seems to be simply unsolvable without creative insight. What is urgently needed, then, is an aha-effect that everyone knows from their own experience:

Every good joke consists of a sensual, figurative contradiction whose solution can neither be rationally explained nor rewritten in words. Nor can the solution to this contradiction be consciously forced or worked out with diligence. Understanding the joke requires a new neural configuration in the brain that abruptly resolves the conflict through a new, clarifying picture. This is the aha-effect, the click in the mind: Only this individual enlightenment enables us to really understand the punch line of a joke. That is real cognition, even in science. So, the worst thing that can be done to prevent the individual and social intellectual progress of humanity is to ignore cognitive dissonances and logical contradictions, to marginalize them, to declare them insoluble or to elevate them to a principle. This is exactly what NIELS BOHR unfortunately did with his interpretation of the quantum problem as "wave-particle duality". No wonder that the *actual punch line of this joke* (EINSTEIN) could not be understood until today.

For this reason, theoretical physics threw in the towel as early as 1927. This explains why official physics stopped searching for a solution to the wave-particle paradox almost a hundred years ago. In order to understand why BOHR and his collegues were so desperate that they considered the greatest cognitive problem of physics to be unsolvable, one should at least know the double-slit experiment and the phenomenon of partial reflection at a simple glass pane. Both experiments are basically so simple that the wave-particle paradox, the deepest problem of theoretical physics, can be easily understood even by layman:

The double-slit experiment

A simple, unassuming experiment that does not even require technology is still today the ultimate test of the explanatory power and validity of physical hypothesis and theories about the nature of reality. The principle is simple. It consists of two holes, matter between them, and sunlight: Sunlight shines through a screen that has two small, closely spaced openings. In the laboratory, an electric light source or laser is used, which emits monochromatic light. When the light propagates through the two openings, a stripe pattern of light appears on a wall or detecting screen behind them (Fig. 1). This can also be a sheet of paper, a photo-sensitive film or a light-sensitive electronic detector. Bright and dark stripes are clearly separated, so that no or hardly any light arrives in the dark areas. However, when one opening is covered, the stripes disappear. Only a normal, uniform light spot then remains behind the other opening (Fig. 2). Now the light can also reach places where no light could arrive at all with two openings, where the dark areas were. That is actually already all – and the deepest enigma of nature. It is not understood until today.

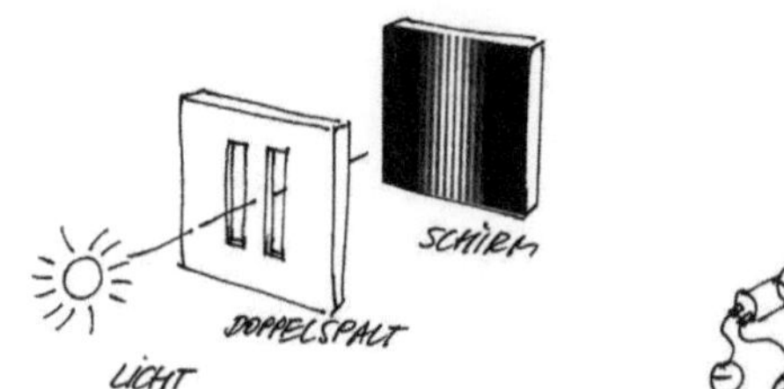

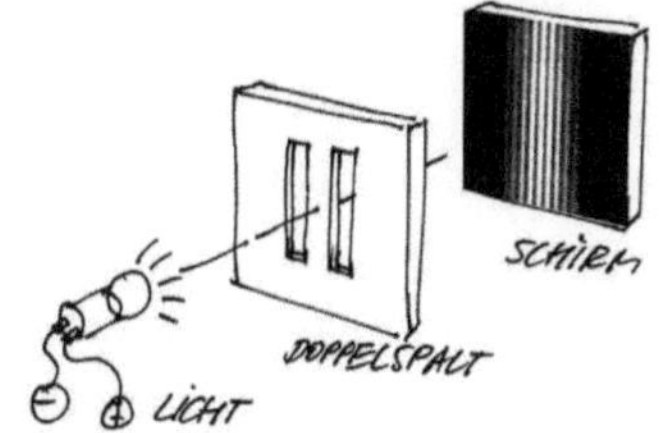

Figure 1 Double-slit experiment with sunlight or electric light

THOMAS YOUNG, who invented this experiment around 1805, came to the conclusion that the sunbeam must split at the double slit into two secondary partial rays, which can subsequently influence each other. This he called *interference*. To explain the stripe pattern, he relied on the wave hypothesis of HUYGENS. According to this theory, light does not consist of particles, as NEWTON had assumed, but of disturbances of an extremely diluted gas which propagate in the gas like ordinary sound and water waves. This hypothetical gas, called *aether*, should fill the universe and represent the primordial matter from which all material structures originate. The two slits act as secondary light sources, from which the two partial waves propagate again in space and superimpose like water or sound waves: In some places they reinforce each other, in others they cancel each other out. This wave model explained the interference stripe pattern so well that the existence of light waves – and different wavelengths – seemed *almost* certain. What was undoubtedly certain, however, was that a *division* of the incident light beam into two partial rays takes place and that these must interfere with each other in order to be able to produce the stripe pattern. This fact alone clearly refutes NEWTON's atomistic particle conception.

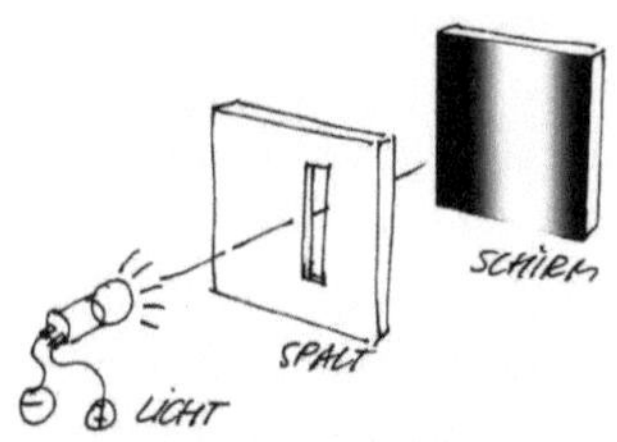

Figure 2 Double slit, one slit closed

Figure 3: The idea of indivisibility fails

The quantum problem is now that the experiment proves that light cannot consist of particles, but quantum 'theory' still pretends that light can be treated *as if it consisted of particles*! The same is claimed for double-slit experiments with matter radiation assumed to consist of particle-like electrons, atoms or molecules. However, a body cannot pass through both openings at the same time without splitting, just as a ball cannot fly through two windows at the same time or a cyclist can drive around an obstacle on both sides simultaneously (Figure 3). A particle would go like a soccer ball sometimes through one, sometimes through the other opening, but never through both simultaneously. If the light consisted of corporeal particles or tiny volume elements of certain energy, only a normal light spot would have to appear behind each opening – at one hole one, at two holes two. Even if the particle could divide, two fragments would be produced, which by definition are bodies again and could not produce a stripe pattern, but only two light spots. However, what is observed is a stripe pattern, which can only be explained by the fact that the radiation passes through both slits simultaneously, superimposing itself, which leads to stripe-like extinction and amplification of the light intensity. For this reason, the nature of radiation can no longer be explained with the body idea of mechanics and the atomic hypothesis. Exactly this had been recognized by THOMAS YOUNG in 1805, at least for the nature of light.

However, one hundred years later, EINSTEIN proposed a new hypothesis that challenged that wave model. In 1905, he suggested that light and electromagnetic fields in general have an as yet undiscovered structure. This meant that light cannot be of continuous nature and light waves or rays cannot act continuously with matter as had been believed until then. This unknown structure causes, according to EINSTEIN, that the light radiation always acts pulse-like at one place only and transfers only a certain amount of energy to the matter at this moment. In the absence of better ideas, he illustrated this effect with particle-like acting energy quantities, which he called *light quanta*. This gave the impression to some physicists that EINSTEIN had created a particle theory of light. In fact, EINSTEIN had only claimed that light interacts *as if* it consisted of particles. So for EINSTEIN the quantum particle was only a provisional auxiliary model, because at that time it was clear to every physicist that indivisible particles are not compatible with the divisibility condition of the double-slit experiment. But only between 1923 and 1926 it became finally clear that even single light quanta, electrons and atoms should show the same branching behavior at the double slit as continuous light beams – at least theoretically, since these were still pure gedankenexperiments (conceptions in the mind). When these experiments were actually carried out many decades later (between 1959 and 2003), these considerations proved to be correct. The fact that even single light quanta, electrons, atoms and even molecules must divide at the double-slit is circumscribed by the term *wave-like behavior*, the fact that the respective emitted amount of radiation energy always acts as an energetic whole at only one point as *particle-like behavior*. This is the deepest problem of physics, even today.

In these experiments, one can also observe that the stripe pattern collapses not only by blocking one of the two holes, but by any energy-transferring interaction with matter – i.e. when matter absorbs light. This can be, if both holes are open, the collecting screen itself (this becomes visible as a light spot), or any opaque object held in one of the two light rays – a cardboard strip, a hand, anything. If, for example, a light-sensitive detector is placed behind a hole, the interference stripe pattern collapses whenever the detector registers light, i.e. consumes radiant energy. The same applies to radiation, which is supposed to consist of electrons, atoms and molecules.

If one reduces the radiation intensity extremely, one can prove that the interference stripe pattern is created from many individual light dots or, in the case of matter, of point-like absorption or registration events of individual electrons, atoms or molecules. This shows that radiation does indeed act locally in a pulse-like manner, consisting of single emission and absorption events, as EINSTEIN had correctly predicted for light. In case of matter radiation this was theoretically no surprise, because matter radiation was assumed to consist of single particles. Only the stubborn interference and divisibility condition, which excluded a particle interpretation of the absorption act, seemed theoretically inexplicable. Thus, the apparently continuous radiation consists in reality of incredibly fast successive individual absorption events. The fact that the energy portion emitted in each case is always registered *as a whole* at only one point on the screen seemed to provide some evidence for the assumption that radiation consists of indivisible energy quanta or particles rather than continuous waves. However, this observation also makes clear that superposition or interference phenomena cannot depend on the intensity (aka brightness) of the light or the density of the respective matter radiation. This fact collides head-on with the atomistic assumption, i.e. with the indivisibility hypothesis, because this means that even single 'particles' and light rays must divide at the double-slit.

Since the interference stripe pattern is similar for light and matter radiation (it differs only in scale, Figure 4), we can conclude that the division process at the double-slit and the holistic action are based on the same physical principles. Although the points of action are distributed randomly across the collecting screen, they are always located in the areas where the stripes form after many registration events. This applies even if the brightness of the light beam or the density of matter radiation is so low that the emission event consists of only one elementary unit (photon, electron, atom or molecule). From this it can only be concluded that even individually emitted light quanta, electrons, atoms or molecules must have crossed both slits *simultaneously.* Thus, they must have divided and formed two partial waves or partial rays, both of which must have contributed to the energetically holistic absorption event. In other words, they must have divided at the double-slit and reunited at the registration point. So, whatever quanta, electrons and atoms really are, so whatever our theory of the constitution of matter and light may look like, it must be able to explain above all this mysterious division process – which excludes the particle concept and the atomic hypothesis on principle. Thus, the experiment shows a *divisibility condition* which is the cause for the bi-parted 'field' state and the resulting distribution pattern, and a *symmetry condition* which ends the bi-parted state by an energetically holistic absorption event.

Note that this is essentially a phenomenal, theory-neutral description of the double-slit experiment, which is basically unanimously accepted. So far, I have actually only spoken abstractly about the

Figure 4 The interference pattern in double-slit experiments with light and matter
The same stripe pattern as with light, only in different sizes, is also shown by double-slit and interference experiments with single electrons, atoms or molecules. In these experiments special detectors are used instead of the collecting screen to register the individual action points. This does not change the principle. You may therefore replace the flashlight in your mind with any 'particle' source. More precisely, with a source capable of emitting light or matter radiation with extremely low density. The lowest possible radiation intensity or gas density consists of individually emitted units that are imagined as indivisible 'particles' (photons, electrons, atoms etc.). But exactly this paradigmatic assumption contradicts the experiment, because the radiation must divide at the double-slit independently of its energy density. This is the so-called divisibility and interference condition, also called superposition principle, which also holds for single 'particles'.

behavior of radiation, which must divide at the double slit and then reunite at some point in order to act energetically as a whole. This physical process must always be the same independent of the energy or matter density of the respective radiation. If I have mentioned terms like photons, electrons and atoms, it is only because these are terms that have already become firmly established among physicists. They appear to them as *necessities of thought* without it being clear what they really mean physically. And this is exactly the problem. The crucial question is how to interpret the experimental facts reasonably and what they actually say about the constitution of matter and light. Opinions differ so much on this question that until today no satisfactory interpretation could be found that is convincing.

Identification of the first suspect: The atom hypothesis

So, behind the double-slit experiment lies the question of the true constitution of light and matter: Do light and matter consist of indivisible particles, divisible waves or of a still *unknown third*, as Einstein already assumed for light in 1909 (Fig. 5). And since 1925 we know definitely that the same question applies to the structure of matter. In other words, for almost a hundred years, we have needed a new physical model that can capture the true constitution of matter and light. But that doesn't seem to be so easy: So far, physicists have failed at it in rows – including EINSTEIN, BOHR, HEISENBERG and SCHRÖDINGER, and all those who came after them. The problem is that this goal cannot be achieved either with wave or particle conceptions nor with a compromise of both: If we assume that light and matter radiation consist of waves, the following question arises: How can we explain that a wave that divides into two partial waves when passing through two slits (Figure 5, center) can later reunite at a certain point on the screen to transmit the whole emitted energy? (Figure 5, right). Conventional wave theory cannot do this, but it is needed to model the simultaneous passage through the slits and the resulting superposition of the two secondary waves that should produce the stripe pattern by cancellation and amplification (Figure 5, center). Even if one wants to save the atomic particle idea as in the quantum mechanical interpretation, the wave model is still needed because one cannot model the interference pattern of action points that forms after many repetitions of the experiment with individually emitted photons, electrons, atoms or molecules without assuming a superposition of two partial rays or waves. Thus, the conclusion cannot be avoided that even individually emitted 'particles' or light pulses (with the minimal quantum of energy) must have divided at the slits. Since they always act as a whole at only one point and not at two (as in Fig. 5 left), the partial waves should actually reunite in the absorption event (Fig. 5 right).

So the crucial question, metaphorically speaking, is what the skier and his tracks mean for the constitution of light and matter (Fig. 6). Does this picture, which experiments literally force upon us, make any physical sense – and if so, of what kind? Is the process of division and reunification real or just fictitious, as the quantum me-

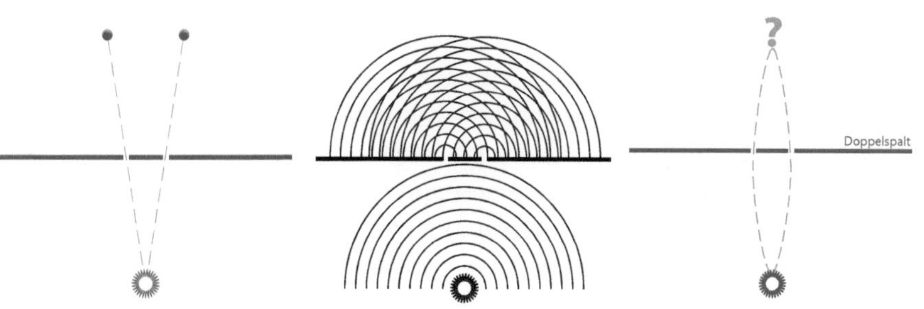

Figure 5 The quantum puzzle at the double-slit
The question of the true nature of light and matter:
Particles, waves or a still unknown third?
left: particle model I center: wave model I right: the still unknown third

chanical interpretation would have us believe? Do these traces symbolize motion tracks in reality, or are they just illusions? Can they be understood as trajectories, i.e. as motion paths of corpuscles, or must they be interpreted as partial waves? But if they are supposed to be waves or rays, as in the theories of YOUNG, FRESNEL, FARADAY and MAXWELL, we have to ask ourselves, why the wave theory can neither understand and represent the physical connection nor the local reunion of the two partial rays. That it cannot do so becomes especially clear in the case of single emitted light quanta,

Figure 6: The quantum puzzle. How can something divide itself, but physically remain a whole ?

electrons or atoms. How the interference stripe pattern then comes about after many trials? If these tracks represent a process in reality, it must be possible to explain how the skier (the so-called "quantum object", i.e. an photon, electron, atom, or molecule) was able to divide and reunite. If they are not supposed to represent a process in reality, it is not possible to explain rationally and causally how the "quantum object" (skier) got from the start point to a registration point at the stripe pattern and what happens in between at a tree, an obstacle or a double-slit. Since it has not been possible to understand that division and reunification process with the then known theories of physics – neither with the atom hypothesis, nor with the wave theory, nor with the gas theory, nor with the special theory of relativity – the quantum mechanical interpretation of NIELS BOHR has retreated to the latter position and denies the reality of particles, waves and of a division process, although the experimental evidence for a real existing radiation and a division process of this radiation is indisputable. This is weird, at least for a natural science, because the interpretation of these experiments is clearly about understanding the true constitution of Nature and their interaction processes.

The main problems were obviously the failure of the atomic hypothesis and the deficiencies of the wave theory. If the atom hypothesis was correct, there should be no division at the double slit. But such a division must undoubtedly take place, which just means that we can no longer speak of an object or particle nature of matter. This is exactly the reason why the wave theory was originally designed. The traditional wave model, on the other hand, is not able to understand the *energetic wholeness* of a bifurcated wave or ray structure, which must be preserved despite the division process. For this reason, this wave model cannot explain the local action of a divided ray system, to which both partial rays must contribute energetically if the principle of conservation of energy is still to apply. Thus the double-slit experiment reveals a contradiction between wave theory, which was designed to reflect the divisibility condition, and the fact that radiation always acts in certain energy amounts at only one point. To solve this contradiction reasonably and to find the true structure of matter and light was the original task of quantum theory, at least for EINSTEIN.

What most physicists consider extremely improbable, if not outrageous and anti-scientific, despite all experimental evidence, is that the atomic worldview of modern physics, based on a 2400-year-old philosophical speculation, could simply be wrong. This idea seems to be unthinkable, because in physics the rumor stubbornly persists that the existence of atoms was unambiguously proven in 1910 (by EINSTEIN and PERRIN, by the way). Even more, despite of *molecular orbital theory,* there is still a misconception that all chemistry proves the correctness of the atomic hypothesis. On the other hand, it was already clear since the introduction of the particle-like electron in 1897 that the atom concept should not be taken too seriously, since atoms were apparently somehow divisible after all. They seemed to consist of smaller constituents, of electrons and protons, which in turn were considered indivisible particles – until one recognized in 1925 that even these 'elementary' particles must be divisible, which was modeled with a wave theory. But as we have just seen, this kind of divisibility cannot be understood with either the atomistic paradigm or traditional wave theories.

However, if we claim to pursue a natural science based on empirical evidence, then we must follow this track. The divisibility condition is such a strong argument that it would easily override the atomic hypothesis, if the quantum mechanical interpretation did not tempt us to continue to believe in a quite nebulous existence of indivisible particles. Quantum mechanics negates their divisibility in a tricky way by claiming that the atomic particle somehow exists and passes through only one slit, but somehow knows whether the other slit is open or closed (because this fact determines whether an interference pattern emerges at all). With this unphysical and most tricky way of speaking, the quantum mechanical interpretation disguises the fact that it is forced to respect the experimentally proven divisibility condition just like any other interpretation. Even in quantum mechanics one is forced to use a wave equation which divides into two partial waves and thus models the simultaneous passage of both slits, because only in this way can the pattern be modeled mathematically correctly. However, in quantum mechanics, the divisibility condition is only indirectly reflected by a mathematically – not physically – introduced superposition *postulate.* Since one believes to have to save the atomistic idea, it is not recognized and interpreted as physical divisibility condition. Therefore the quantum mechanic, unimpressed by the experiment, continues to speak of indivisible particles and persuades himself that light quanta, electrons and atoms double at the double slit or at an obstacle in an esoteric-magical-virtual way – instead of *physically* dividing!

This illogical way of speaking and thinking shows crystal clear that the proponents of the quantum mechanical interpretation have not yet grasped the punch line. Since no one had any idea why the atom hypothesis failed completely and the wave theory failed at least partially against reality, many physicists began to accept the Copenhagen Interpretation with a shrug – even though it didn't seem to make any physical sense at all. The Copenhagen Interpretation, i.e. the quantum mechanical interpretation by NIELS BOHR, saved the muddled situation by claiming that the atom hypothesis, the body model of mechanics and the wave theory fail only at the deepest level of nature, but otherwise remain valid – or at least useful and indispensable for the scientific method. In this way, BOHR separated nature into a submicroscopic and a macroscopic level, for which different physical laws should apply. At the submicroscopic level, BOHR claimed, there simply no longer existed an *objective reality* that we could describe with physical models. So if one wants to summarize BOHR's strange philosophy in a few words, his message was:

To save the atomic paradigm, you must deny reality.

Of course, if you are young, unprejudiced and not yet overloaded with superfluous erudition, and look at the key experiments from a purely geometrical point of view, you will probably think (as I did, when I was just 17): That's weird. Were physicists really so desperate that they had to accept such abstruse conclusions? Shouldn't a reasonable scientific conclusion express the opposite? Like this:

If the experiment disproves a hypothesis, not the reality concept is in danger. So the double-slit experiment only shows that the atomic hypothesis must be wrong.

If one has understood the problem even halfway, one is stunned and wonders: Has nobody really recognized so far that we must have to do here with a branching process – like everywhere in nature? Is it not obvious that the atomic hypothesis fails in these experiments, as well as the body concept of Newtonian mechanics and the wave theory in its present form? Wouldn't it be the most normal thing in the world to accept this fact unconditionally, solve the puzzle and design a completely new model of reality? Wasn't it clear that now all contemporary models and theories had to be questioned, especially the *atomos* hypothesis? Why has physics not taken this way?

Obviously, it's all about the right epistemological stance:
It all starts with taking experiment seriously as the judge of physical ideas and theories and not suppressing the failure of the atomic hypothesis. So if we acknowledge without further ado that these experiments prove that the atomic concept of matter and light already fails at the double-slit (and likewise for light at beam splitters, glass panes and polfilters), and that the energy portion emitted in each case always comes into effect locally as a whole, despite the simultaneous passage of the double-slit, the real physical question arises:

How can something divide itself and still remain a whole?

This is what the experiments actually show, and the very question that so embarrasses modern physics (Figure 6). As Richard FEYN-MAN put it:

"The two-slit experiment contains the sole mystery of quantum theory [...] How does it work? What is the machinery behind this law? [...] No one has found any machinery behind the law. No one can explain any more than we have just explained. [...] We have no ideas about a more basic mechanism from which these results can be deduced."

However, the last sentence is no longer true, because we now have an idea: The crucial point is to understand the true nature of the division process, which can no longer be understood as a *mechanical* division. It follows that this 'mechanism' must be holistic in nature, i.e. a genuine physical process in which light and matter can divide holistically. This is a real discovery. It is based on the experimentally proven fact that the division at the double-slit cannot be mechanical in nature, since no separated parts are created in the process (if it were, there would be no interference stripe pattern). So it is not a splitting or *breaking into pieces*, as in the mechanical worldview. Therefore, the only logical conclusion is that light and matter rays or waves can branch like trees at the slits and thus remain a *physical whole*. The local energetic symmetry condition then shows that this division process must be reversible in the case of an absorption event. The picture that then unfolds before our mind's eye shows a branching process in which a single ray of light divides. If we imagine that light propagation takes the form of a spherical wave front, we get a picture in which the light sphere divides just like a biological cell. The same picture applies to spherically expanding matter radiation, whose division immediately reminds us of the formation of a diatomic molecule. This branched structure is probably exactly the field structure for which EINSTEIN searched in vain.

Remember: Once you eliminate the impossible, whatever remains, however improbable, must be the truth.

This is the logical chain of thought that most quantum physicists have been trying to avoid to this day, because it seemed unthinkable and completely impossible that the atom hypothesis – the indivisibility assumption – could be *fundamentally* wrong. Although such a conclusion had to be drawn already two hundred years ago, when it was about the nature of light, physics refuses until today to draw the same conclusion concerning the nature of matter. So the unthinkable is the experimentally well-founded insight that the assumption of indivisibility cannot apply to nature and reality in general. This is real cognition, a click in the mind.

And thus the first suspect in this tricky quantum case is clearly identified: The impossible is the atom hypothesis, the 2400-year-old philosophical assumption that matter consists of indivisible particles. So our most important and widely believed scientific hypothesis must be plain wrong. This may be hard to believe, but it is exactly what the double-slit experiment proves.

Identification of the second suspect: The division process in wave theory
One might now think: If the atomistic particle concept is so clearly refuted by the experiment, the wave theory must be correct. After all, YOUNG successfully replaced the atomic particle model of light by the wave model, which is still valid today – at least for classical optics. In a special form, the wave model even applies to the quantum physical description of matter, which in principle behaves like light at the double slit. But YOUNG did not yet know what we know today: That the apparently continuously produced interference pattern is generated by incredibly rapid successive point-like absorption events. That was EINSTEIN's really ingenious insight – and this at a time when experiments with single photons, electrons and atoms were not yet possible at all and only the photoelectric effect pointed to some inexplicable phenomena. So every energy transfer, every energetically *effective* action consists of a local absorption event, in which the entire radiated energy portion must come into effect as a whole – despite division at the double-slit. And with this, also the wave theory breaks down, because it cannot explain how and why the branched and spatially distributed radiation should reunite at one point. This problem has already a spherical wave, but becomes much more clearer in partial reflection and double-slit experiments with single light rays or light quanta. This means that the continuous wave theory is energetically and temporally not reversible. So we have to assume that the theoretical claims and the practical capabilities of wave theory diverge at some point. And this divergence reminds us of what is really at stake: Does wave theory even know real branching processes? Did the idea of a *physical branching process*, which automatically implies the principle of wholeness, so to speak, even exist at that time?

At first glance, it looks like optics – and the associated wave model – know branching processes very well. After all, the wave model was developed by YOUNG (1805), FRESNEL (1817) and MAXWELL (1864) specifically for the purpose of explaining the splitting and bending of light beams by glass bodies, obstacles and slits. The observed diffraction, superposition and interference effects were then explained by the idea that the spherical wave fronts and focused light beams could be wave-like deformations and energetic excitations of a yet unknown gas with extremely low density. Therefore, the propagation of light was assumed to be similar to the propagation of sound waves – although the carrier medium of light was now more like a vacuum with negligible gas residues. However, between FRESNEL's optical wave theory and the electromagnetic wave theory lies a decisive step: The terms wave-like deformation and energetic excitation now hide *electric* and *magnetic polarization processes* which FRESNEL did not know yet. MAXWELL had integrated them into the wave theory, but could never clarify their true physical nature. Although this has not changed until today, this wave model is still valid – unless we want to understand the behavior of individual light rays, which is basically the same as that of single light quanta.

So, at second glance, it turns out that the wave theory can describe the branching of light only phenomenologically, but not physically. This is because real branching processes were not considered at that time (which has remained so until today). This is due to a composite-hypothesis, which is based on the assumption that light beams always consist of many individual light rays, which at this moment only separate spatially, i.e. split mechanically. That is, they were and remain individual, separated components. This idea is now joined by the problem of polarization. The theory assumes that light waves are always *transverse* waves, which means that they should already have a definitive plane of polarization in space and that they are polarized *from birth*. With this assumption FRESNEL had explained polarization: If a light beam branches at an obstacle into two partial rays, which then turn out to be polarized in opposite directions, the individual light rays must have been sorted somehow according to their polarization into the new directions. Apart from the question how this sorting should work exactly, the real problem arises if one wants to imagine only one single ray: How can something divide into two parts without presupposing the existence of two components? This is the dilemma of mechanics, molecular theory, and all composite hypotheses, thus also the theory of light.

To explain the branching of a single ray with the wave theory, one would need at least two transversal waves in the incident light ray, which would already have to have the exactly opposite polarization.

However, the polarization of the resulting partial rays can only be determined after the bifurcation, so that the presupposition remains unprovable. In fact, this presupposition proved to be untenable in quantum theory and led to long-lasting irritations: Apparently, the polarization is not already present, but is only generated in the process of branching! Since branching is excluded a priori, wave theory cannot explain the behavior of a single light ray or light pulse, the resulting opposite polarization of the partial rays, and the conservation of energy in the absorption event. This seems surprising at first, because the wave theory knew already two hundred years ago that the behavior of light beams in optical experiments cannot depend on the intensity or brightness of the light, so it should be valid also for single light rays with minimum energy. But this is not the case, because of the definition of polarization. With this insight, the reality problem of quantum waves also takes clearer shape. It sheds new light on the design problem of the theory and shows that the nature of polarization has remained unclear to this day. Since electric and magnetic polarization processes form the basis of MAXWELL's wave theory, the electromagnetic wave theory stands on shaky legs as long as the true nature of polarization is not clarified. The same problem shakes the quantum theory of light and matter. In short, we need to understand that *polarization* and *branching* are the same thing.

Thus we have identified our second main suspect in this mysterious quantum case: It is wave theory. It fails for single emission and absorption events, because holistic division and branching processes are as foreign to wave theory as they are to the atom hypothesis and to classical mechanics.

Obviously, we look at the world through the wrong glasses. In wave theory, this deficit can be traced back to a hidden assumption that has been overlooked so far: That single light rays or waves cannot divide either, just like atoms! This premise prevents the modeling of emission, branching and absorption of single light rays on principle, but is clearly refuted by quantum double-slit experiments. So the composite-hypothesis plays the same cognitive blocking role in the wave theory of light as the atomic paradigm in the theory of matter. This seems to be the second reason why the quantum puzzle seems unsolvable until today. But that's not all:

Identification of the third suspect: The kinetic gas theory of mechanics

It is particularly strange that the wave model must also be used to describe the behavior of single electrons, atoms and molecules. As we have just seen, the true reason is the divisibility condition, the simultaneous passage of the double slit or of two possible directions. But wave theory does not really understand the nature of this division (incidentally, modeling bifurcation processes requires nonlinear mathematics and nonlinear complex dynamics, though this was only discovered between 1975 and 1985). However, the divisibility condition cannot be ignored, because it is an experimentally proven fact that characterizes the most fundamental physical property of matter and light: *Holistic divisibility.* The double-slit experiment also proves that essentially no physical-ontological difference exists between light and matter, at least not as far as *gaseous matter* is concerned. This is a decisive hint which will still play an important role: *Only if the matter is vaporized,* i.e. brought into a gaseous state, matter-gas beams in the experiment behave like light beams, passing through both openings simultaneously and forming two partial rays that somehow interact with each other and form, in principle, the same stripe pattern. In other words: matter radiation is nothing else than an emitted gas with extremely low density. Doesn't that sound familiar?

In MAXWELL's wave theory, the propagation of light is conceived as an electric and magnetic polarization process of an extremely rarefied gas, which causes highly elastic deformations in this gas (since they are completely reversible) that propagate in waves. Thereby it is presupposed that the entire gas volume is in a state of rest and a wave represents only a local and temporary deformation of the gas. When we speak of matter radiation, the radiation is also a gas, but a known one, which behaves in the same way as light at the double slit: Consequently, the gas blow or jet must divide (holistically) at the double slit, regardless of the gas density. So, in the extreme case, we must imagine a gas volume consisting of only one atom or only *one volume element with mass*, which can divide at the double slit and reunite on a collecting screen. However, this conclusion from experiment clashes with two powerful theories that seem to be so well supported experimentally that they cannot be challenged without shaking the world view of entire branches of physics:

One is the kinetic theory of gases of physics, the other the molecular theory of gases of chemistry. The former assumes that gases consist of tiny particles of mass that are constantly buzzing around like a swarm of mosquitoes (in a vacuum!). The second one says that these gas particles are either indivisible on principle (called atoms), or at least not divisible under these circumstances (called molecules). The next logical problem is that in such a theory one can hardly consider a particle as a gas: If one assumes that this particle has an invariable, extremely tiny size and is supposed to be structurally stable like a billiard ball, one can no longer attribute to it the properties of a gas such as pressure, temperature, volume expansion or change of aggregate state – let alone explain the division, propagation and superposition at the double slit (which now resembles the phenomenon of turbulence). This is followed by the next problem: In particular, this assumption rules out the possibility that a tiny particle can expand to such an extent that it can pass through both slits simultaneously. Because for atoms and similarly tiny 'particles', the two slits have a distance that can be 10,000 times their own diameter and more (if the size of the atoms were invariant). This is basically also true for molecules, since it is assumed that molecules are stable in the gaseous state and do not decay. Therefore, it is assumed that gas particles cannot divide at the double slit. In chemical reactions, of course, the situation is quite different: In this case, it is absolutely clear that molecules must divide. Whether we can explain it offhand or not – the double-slit experiment requires that 'particles' or *volume elements with mass of whatever density* are divisible, as long as they represent fluid or gaseous states. From this follows that the elementary structures of light and matter must be holistically divisible in the same way.

But why can't gas theory explain holistic division and branching processes? Why has no one been able to recognize holistic division processes in gas and molecular physics – although they are so obvious in the experiments of quantum physics? The answer lies buried under fundamental assumptions (now in chemistry) that common sense takes as almost self-evident and proven beyond doubt. Apparently, these premises have never been critically questioned, despite the obvious contradictions in the double-slit experiment.

If we take the experiments of quantum physics and the associated problems of cognition seriously, the particle assumption simply cannot be correct. It follows that we have to critically re-examine the molecular gas theory of chemistry and its genesis. In the second half of the 19th century, the kinetic theory of gases was added, which was developed by physicists. One of them, by the way, was MAX-WELL. In this context, it is important to mention that the kinetic gas theory is at the same time the basis of the heat theory, because in the mechanical world view, the thermal energy of a substance is equated with the kinetic energy of moving particles. That this conception must be wrong, if the particle conception is wrong, should be obvious. Thus, the undoubtedly coherent findings concerning the 'degrees of freedom' of the molecular structure must be explained in a different way. At the end of the century, a new generation further developed the kinetic gas theory into thermodynamics by trying to combine the properties of gases and solid matter with the properties of electromagnetic radiation. But the still immature physical ideas about the nature of light and matter thwarted this program, which led via PLANCK and EINSTEIN to the birth of quantum theory.

When physicists tried to understand these contradictions in the first quarter of the 20th century, hardly anyone thought of going back a hundred years and questioning the basic assumptions of gas and wave theory once again. When they thought about matter radiation and its strange behavior in then still hypothetical double-slit experiments with particles (and in similar but real diffraction experiments with beams of electrons), they were initially unaware that they were presupposing the validity of both the gas theory of chemistry and the kinetic gas theory of mechanics. Again, EINSTEIN was the first to notice this and to begin work on a quantum theory of gaseous and solid matter. However, a new, non-mechanical gas theory never emerged from this, although it was definitely in the air, and still is.

If, after all these experiments could be carried out in reality during the last thirty years, we still want to imagine that gases, solid matter and light consist of indivisible particles, then we are really naive. Because then we apply the atomic hypothesis, the classical molecule theory and the kinetic gas theory of mechanics to quantum physics in a way that lacks any critical thinking and historical reflection. In view of the divisibility condition of the double-slit experiment, such an attitude can only be called negligent or paradigmatic.

And thus the third suspect in our tricky quantum case is identified: It is the gas theory. The testimonies of the kinetic gas theory and the molecular theory do not agree with the double-slit experiment in crucial respects.

Thus, we need to develop a new kinetic gas theory and a new molecular theory that can model holistic division and branching processes. By the term 'kinetics' we then understand not only the changes of location of structural invariant material objects caused by forces, i.e. the motion and acceleration of bodies in the sense of mechanics, but also those branching movements of matter which represent structural changes and involve – or generate – electric and magnetic forces. And that leaves only one suspect, hopefully the last one:

Identification of the fourth suspect: The definition of the speed of light
Equally disturbing and confusing for our current understanding of Nature (reality) is that experiments with branched rays prove that there must be holistic physical connections in the branched ray system that can act instantaneously over arbitrary distances, as if time and distances do not exist at all. Since light can propagate arbitrarily long and far (if it is not absorbed by matter), the two branches could in principle extend through the entire universe. This phenomenon is called *non-locality*, but more appropriate would be idea of a *holistic field structure*, provided, of course, that this field state exists in reality – which the experiments more than suggest. The same applies to branched matter radiation. The problem with the speed of light becomes particularly clear when a light beam splits at a beam splitter (or for other reasons) in such a way that two partial beams (B_1 and B_2) move away in opposite directions (Figure 7). If after some time, or at any distance, a determination of the magnetic polarity is performed on the partial beam B_1 (a so-called spin measurement), a certain 'direction' of the magnetic polarization is obtained without being able to know beforehand in which 'space direction' the

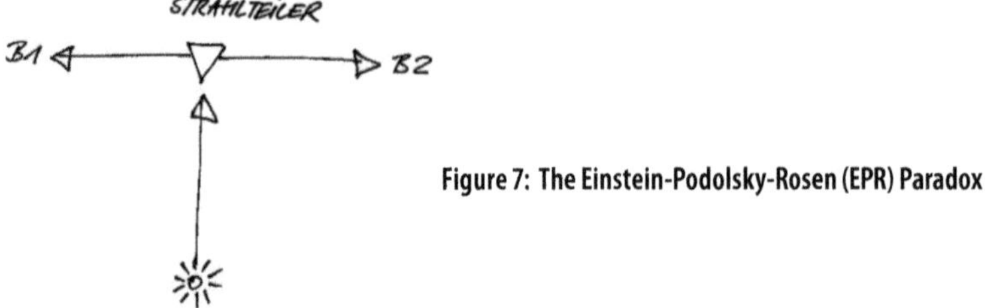

Figure 7: The Einstein-Podolsky-Rosen (EPR) Paradox

Spooky action at a distance? Virtual particles? Probability waves? A reality that cannot be understood by ordinary mortals on principle? I think there is only one reasonable explanation: We are dealing with a branching process of single light or matter rays, which creates an opposite magnetic polarization of the partial rays. So what we see is the formation of an enantiomorphic field with magnetic dipole character. This is the true nature of polarization, a structure forming process that electromagnetic theory cannot map and understand. On the one hand probably because Hertz modified Maxwell's equations, on the other hand because the unidirectional definition of the speed of light makes such an interpretation impossible.

magnetic 'field vector' will point. At this moment it is certain that the other partial beam B2 must have exactly the opposite magnetic polarization. This is valid again independently of the energy density, thus also for single branched light rays, light quanta and likewise for matter radiation, which is assumed to consist of single electrons, atoms, etc. If one insists that these partial rays do not exist in reality, but represent only 'probability waves' (as the quantum mechanical interpretation does), one is forced to assume – as in the case of the double slit and partial reflection – that the original quantum particle splits 'virtually' and takes both ways at the same time. So we have to do again with the same reality problem as in the double-slit experiment. The quantum mechanical interpretation describes this state with a wave equation which says that the system is in a superposition state in which two oppositely polarized 'virtual' quantum particles 'exist' simultaneously (in the mind, but not in reality). At this point, it is again easy to see that we are dealing with a branching process that generates an enantiomorphic state with opposite magnetic polarization. However, three non-experimental things prevent such a reality model: The atomistic paradigm, the indivisibility of single waves and light rays in the transverse wave (aka polarization) theory, and EINSTEIN's theory of special relativity.

For the quantum mechanic, however, this does not look so simple as long as he imagines particles: Since the virtual particle whose spin direction is determined cannot have had this property before, this property must have been generated by the measurement itself. But this would mean that this spin measurement must have transferred itself instantaneously, without time delay, mirror-symmetrically to the other, arbitrarily far away particle.

Interestingly, it was EINSTEIN, of all people, who discovered these strange *nonlocal* aspects of quantum physics in 1935, which pointed to a physical connection between two particles or two real existing partial waves or rays. Since he still held to his provisional particle picture of the quantum and could not doubt the existence of particle-like electrons either (although he had helped deBROGLIE's wave theory of the electron to a breakthrough), he called this connection a *spooky action at a distance* and concluded that the quantum mechanical interpretation of BOHR and HEISENBERG, which seemed to revolve around the strange behavior of real electron particles, must be incomplete. However, it never occurred to him that his own special theory of relativity might be wrong or in need of improvement. Surprisingly, his colleagues never considered this possibility either, for the existence of physically connected (aka holistic) states of arbitrarily large extension contradicted the assumption of *separability* that followed from EINSTEIN's interpretation of the relativity principle. This assumption was based on the *axiomatic postulate* that the speed of light is always constant and a limiting speed that no material body can reach or exceed. Consequently, nothing can move faster than light, not even a physical action. These conclusions were in turn based on MAXWELL's assumption that the propagation of light is a unidirectional motion. In other words: After thirty years, special relativity had become a seemingly incontestable theory.

The debate about what these nonlocal effects mean physically and whether they prove an incompleteness of the quantum mechanical formalism became known as EINSTEIN-PODOLSKY-ROSEN paradox, abbreviated EPR. By the way, both EINSTEIN and BOHR imagined this problem originally as if it were about two electrons moving in opposite directions after a collision and therefore must be subject to the law of *conservation of momentum*. So originally it was not about the magnetic spin of electrons or photons at all; this point of view was introduced only in 1951 by David Bohm. Einstein wanted to refute Heisenberg's uncertainty principle, according to which the usual rules of mechanics should apply to quantum particles only in a limited way. Their long-lasting debate about the nature of reality shows why and how Einstein, Bohr and Heisenberg talked past each other – so much that they could never find a common platform of understanding, even though all three were thinking of real existing particles (but never thought of ray splitting, let alone bifurcation of waves, rays or fields). Interestingly, non of them found any reason to question special relativity, the underlying concept of mechanics, the motion of light, the notion of particles, or the atomic hypothesis itself. And so it has remained until today. And whoever still dares to question the definition of the speed of light a hundred years later must have extraordinary courage, because what was still a postulate with Einstein in 1905, was declared a law of Nature in 1983!

So if one questions the definition of the speed of light today, one is not only attacking a sacred law of nature, but also Einstein's special relativity. This, of course, scares off most physicists. But not all: The first to clearly recognize that there might be something wrong with the definition of the speed of light and the conclusions derived from it was John Bell in 1964. He designed a mathematical relationship that allowed physicists to experimentally prove whether this violation of special relativity actually exists, and whether the quantum mechanical formalism agrees better with the result of polarization measurements than a formalism based on the assumption that the definitive polarization state exists before the measurement. Since experiments with the required precision could not be performed until 1980, the verifications dragged on until 2007 (and are performed regularly to this day). The result: These violations exist indeed, both for light and matter radiation with lowest possible energy density. This in turn was taken as a confirmation of the quantum mechanical particle interpretation of the mathematical formalism, and this as a refutation of Einstein's epistemological claim that there must be an objective reality which physics has to recognize and adequately model. Since it seems impossible to question the special theory of relativity, one speaks only of *correlations* between polarization measurements (either on virtual quantum particles or partial waves) and hastens to assure that in this way physical effects can be transferred in no case. In other words, these 'correlations' clearly contradict the "spirit of relativity" (R. PENROSE 1989), but have no physical effects and cannot be used for faster-than-light communication – at least according to the current state of interpretation.

When in 1993 it became clear, at least theoretically, that EINSTEIN's "spooky actions" over arbitrarily distances must actually exist and consequently cannot be just "mathematical" correlations, the term *teleportation*, inspired by science fiction movies, emerged. Already four years later, in 1997, this experiment could be realized with light quanta. However, this is exactly what EINSTEIN's special relativity rules out, because no real physical effect can propagate faster than light. Since the distance between the two 'virtual' light quanta – or the tips of our branched partial beams B_1 and B_2 – is 2c (both move with the velocity c in opposite directions), only a double speed of light could bridge the distance between the two 'virtual' quantum particles, ray tips or wave fronts. However, since 1997 it is clear that this effect exists in reality. Of course, this not only contradicts the spirit of relativity – it directly challenges it. However, no physicist may seriously consider this: he would first have to design a new theory of relativity and overturn the 'natural law' of a constant velocity of light. Accordingly, the physicists formulate their statements very carefully and speak only of instantaneous transmissions of polarization states which cannot be used for instantaneous message transmissions.

For physicists who have never thought about these contradictions, the attempt to question special relativity (like the atom hypothesis) may appear at first sight as heresy. But, of course, in science there are neither saints nor sanctuaries, but only assumptions, hypotheses, conceptions, ideas, and models about the nature of reality that can be either correct (appropriate) or incorrect (inappropriate). Non-locality appears so puzzling because the experiment collides with the indivisibility assumption, which applies to fictitious quantum particles as well as to individual light rays. In his special theory of relativity, EINSTEIN interprets FRESNEL's transverse wave theory in the electromagnetic garb of MAXWELL, but apparently overlooks the fact that individual light rays are considered indivisible only because of FRESNEL's definition of polarization, which led to the transverse wave theory. So, one could have also questioned the transversal wave concept and thus the nature of polarization. But in the meantime it seemed to be clearly confirmed by the experiments of HERTZ. However, since the MAXWELL-HERTZ-LORENTZ theory failed in explaining the photoelectric effect and local absorption events, but no error in the wave theory was apparent, EINSTEIN looked for another explanation. He started from energy conservation and entropy principles and PLANCK's energy quanta, assumed a still unknown structure of radiation and postulated local energetic-holistic actions of light rays or radiation. He interpreted this *energetic wholeness* as atomic, i.e., as indivisibility, and illustrated his considerations provisionally with indivisible light particles – knowing well that he ignores the division at the double slit, i.e., the interference condition.

So polarization plays no role in either special relativity or the quantum theory of light. Since the assumption of indivisibility of single 'quantum particles' as well as of single light rays must be wrong, as quantum experiments clearly prove today, polarization can no longer be interpreted as decomposition and separation of partial rays. In addition to this, spatial directions have no physical meaning, while left-and right-handed systems and mirror symmetries have one, and a relative one at that. This is the point where holistic divisibility, branching processes and enantiomorphic fields come into play:

The discovery that we can easily understand the physical problems of quantum theory by throwing overboard the atomic idea and the hidden indivisibility assumption of the transversal wave model leads almost automatically to the idea of holistic division and branching processes and a magnetic dipole formation. But this experimentally well-founded physical idea cannot be reconciled with special relativity, at least that's how it seems at first glance. If it were really so, one could close the book at this point and say: It was a brave attempt, but it was clear from the beginning that this idea cannot work. But the actual problem is that the theory glasses of the transversal wave model simply do not allow the wearer to see the reality in this way. Special relativity is no different, since it is still based on that wave theory. The problem is simply that all experiments clearly indicate a branching process that creates coherent (i.e., physically connected) branches with opposite magnetic polarization. At this point, all of physics is perplexed, since none of its theories can map this, let alone understand it. Neither NEWTON's mechanics, nor classical molecular theory, nor MAXWELL's field theory, nor EINSTEIN's special theory of relativity, nor quantum theory in its present form. All five, but not the experiment, exclude holistic division and branching processes from the outset, simply because they do not yet know such a 'mechanism'. So to understand this mechanism of nature, we need a revolutionary new physical understanding. This is exactly what one should expect from a genuine quantum theory, at least according to EINSTEIN.

The only thing that prevents this is, of all things, EINSTEIN's special relativity, which precludes the idea of holistic division and branching processes and therewith a new understanding of the nature of polarization from the outset. At least that is what almost all physicists believe, I suspect. However, to elevate a theory above the experiment without thoroughly checking the underlying assumptions, which are just questioned by the experiment, is of course a scientific methodological error. In fact, we are dealing here with a stack of assumptions that we now know cannot be correct. And this seems to include the concept and definition of the speed of light. At the same time, the relative principle seems to be indispensable. So the question is rather whether we can reconcile our branching hypothesis with relativity.

I think this is indeed possible: if one looks a little closer at the history of science, one finds that the definition of the speed of light is based on the premise that the propagation of light can be understood as unidirectional motion (this is how any motion is defined in mechanics). EINSTEIN's formal resolution of the conflict between the induction experiment and the theoretical treatment in the theory of Maxwell does not challenge this premise at all. Thus, the idea of a unidirectional motion of light is another fundamental assumption that no physicist has ever questioned (understandably, since it is now considered a law of nature). However, if the nature of light is

based on *dipolarizing* field branching processes, we are dealing with
a non-mechanical motion in the sense of structural changes. If this
is really the case, the unidirectional definition of the speed of light
(called velocity) cannot be correct. Theoretically, it should then be
possible to derive the speed of light (or the constant c, whatever it
may mean) from field branching processes, and to explain FARADAY
induction in terms of polarizing branching processes caused only by
changes in distance, i.e. relative motion. And this is nothing more
than the missing non-mechanical interpretation of the dynamic
polarization movement that MAXWELL had originally searched for
but never imagined. So if we want to understand how the definition of the speed of light came about in the first place and how the
deficits of the old theories are to be understood, we have to go back
again into the history of physics and look for their roots.

**And with these considerations, we have identified our fourth suspect in this
tricky quantum case: It is the unidirectional definition of the speed of light,
which seems highly suspect with respect to the quantum physical branching
processes in double-slit, partial reflection, and polarization experiments.**

Solving the quantum puzzle: Almost too much to ask, but not impossible

NIELS BOHR once said: Anyone who is not shocked by quantum
mechanics has not understood it. Well, I am shocked – and maybe
you are too: Our four suspects represent virtually everything what
we have learned in physics and chemistry: the atomic idea, the additive, mechanistic molecule model, the kinetic theory of gases (which
is just another name for the mechanical theory of heat), the wave
theory, and the definition of the speed of light. Now you can perhaps understand how physicists must have felt around 1927 – as if
the carpet had been pulled out from under their feet... Of course,
the achievements of science and technology in the 19th and 20th
centuries would have been completely unthinkable without the classical models of physics and chemistry. Just think of steam engines,
combustion motors, electrical engineering, electromagnetic motors,
steam turbines and power plants, electricity grids, telescopes, microscopes, radio and television, automobiles, airplanes, rockets, and the

chemical industry. Without the fundamental insights of physics and chemistry, such a development of groundbreaking technologies and industries would have been completely impossible. Accordingly, the confusion was great: How could physics and chemistry be so successful if their fundamental assumptions were not correct?

Understandable or not, quantum physics tells us straight to our faces that we have no suitable idea of how nature is really constituted and what is actually going on in reality, especially at the double-slit and in interference experiments. Yet classical physics shows us that we must have grasped many things essentially correctly, even if the fundamental assumptions should not be correct.

One can deal with this fact in two ways: If one finds oneself completely unable to doubt the old, seemingly proven models of reality and one's own wisdom, the only thing left to do is to despair of the real existence of the world and its recognizability. This is the strange 'epistemological' path that NIELS BOHR took and that many physicists have followed, albeit more or less reluctantly. One is then so convinced of the usefulness of the old theories and mathematical descriptions, that one considers nature, which simply refuses to submit to these theories, to be crazy, bizarre and unrecognizable. But one can also look at the whole thing from a self-critical perspective like EINSTEIN – then the story looks a bit different:

At the beginning of the 20th century, Mother Nature gave physics a resounding slap that still burns mightily on the cheek today: It was an unmistakable call to adapt the old world view of physics – and its theories – to reality and technical progress. Many physicists have not seriously considered this challenge until today, although the experiments clearly show that our models of the nature of matter and light must be wrong on crucial points. This seems to concern also the light speed concept as well, which worried EINSTEIN early on. In practical terms, this would mean that physics, the basis of all our natural sciences, is unable to understand the true constitution of nature. This is, of course, a bitter truth, but not really news (think of SOCRATES). However, quantum physicists find it difficult to admit this openly, and this is the real shock that should make us think.

Instead, a lot of physicists believe that they have *"already understood ninety percent of nature"*. But even this overconfidence is not really new: The same was believed at he end of the 19th century before PLANCK and EINSTEIN discovered the new. Have we learned nothing from the evolution of physics?

Can it really be that the assessments of the state of science is so far apart and depends only on the interpretation of the wave-quantum paradox? One thing is certain: The simplified atomistic-mechanistic world view that is still taught today at all schools and universities has been completely outdated for almost hundred years. Of course, quantum physicists know this best themselves and would probably like to change it, but so far they have neither found the right words nor a reasonable explanation. They simply do not yet have a new, meaningful design draft that they could counter the old world view of physics. This is why most books – that are supposed to explain quantum physics – appear correspondingly mystical and incomprehensible. They mostly draw a picture of the quantum mechanical interpretation of quantum physics, which denies a deeper physical reality and its recognizability. And in doing so, these books convey a *surreal understanding of science* in which a negative bias against new and deep insights prevails. You don't believe me? Then listen to this:

"It is wrong to think that the task of physics is to find out how nature is constituted. The task of physics is rather to find out what we can say about nature."[5]

Or this: *"Be content to accept the observable experimental facts as they are. Don't ask why they are so – you would only run into a dead end and be frustrated... So the only thing left for us to do is to describe how events seem to happen in nature. That's all we can accomplish."*[6]

This is how the spirit of BOHR speaks. From the bottom of my heart I can only say: In such an epistemological climate, one cannot make physics. This absurd 'pedagogical' advice prohibits thinking and is therefore unacceptable. Just think about it a little: If man had ever

5   Niels Bohr
6   Tony Heyl; Patrick Walters: „The Quantum Universe". Spektrum Verlag Heidelberg, 1998 (German)

adhered to this maxim, water and ice would still be fundamentally different things... In short, science would not exist at all.

Such a negative epistemological attitude, which flatly denies a solution of the wave-quantum paradox and thus the discovery of what actually happens at the double slit, contradicts everything natural science stands for. If one has understood this, one also understands EINSTEIN's resistance. It was not directed against quantum physics, but against the quantum mechanical interpretation of the *unsolved problems* of quantum theory and its epistemological stance. So how could it come so far that a science with the highest demands on itself develops such unscientific airs and graces? Why could EINSTEIN's plausible point of view not prevail?

The main reason, of course, is that neither EINSTEIN nor anyone else could solve the puzzle. This is not a problem, you may think now, because in the meantime we have produced myriads of intelligent young physicists, one of whom will surely solve the quantum problem. So it should be, but we wait for it already for hundred years in vain.[7] The real reason, as we have seen above, is that the solution of the quantum puzzle is considered impossible (and unnecessary) by the interpretation of BOHR, HEISENBERG & BORN, later also called the Copenhagen Interpretation or Quantum Mechanics for short.

But if one takes the wave-quantum paradox put into the world by EINSTEIN physically seriously and shares his quite reasonable epistemological attitude (without becoming uncritical of his theories), then a thorough analysis of the fundamental assumptions of physics is actually obvious. As we have seen in our analysis of the four suspects, this results as if by itself in a research program which already contains the approaches to the solution.

So why has such a survey never been carried out systematically in physics? This becomes understandable only if one knows the epistemological climate that began to develop in physics in 1927 with the

7 In Germany alone, about 16,000 students enroll in physics every year; worldwide, there are probably at least 200,000 per year... After ten years, that makes an army of 2,000,000 physicists. And among them, there are supposed to be no more Einsteins?

Copenhagen Interpretation. It was a kind of ice age in which critical and creative thinking froze and *"the dark age of theoretical physics"* began (as CARVER MEAD once put it). Of course, such a radical view is only the antithesis of the picture that quantum mechanics likes to paint of itself. Many authors start their books and articles with phrases like *"qantum mechanics is the most successful scientific theory of all times"* to signal that they submit to the prevailing ideology, before they allow themselves to criticize quantum mechanics, which has no intention of producing any physical meaning at all.

Isn't that a bit childish? After all, science truly flourishes only in astutely conducted intellectual disputes. Proclamations of loyalty and the mere invocation of famous names do not contribute to scientific progress...

The fact is that since BOHR's declaration of the unsolvability of the quantum puzzle we have been dealing with a socially agreed cognitive barrier, a convention that declares the individual blockade of thinking to be a kind of *sound barrier of cognition* (at least for the genus of physicists). This convention, in force since 1927, is responsible for the negative epistemological climate in 20th century physics. It claims that human thought is simply incapable of grasping the true constitution of nature and reality because it relies on physical and spatial ideas with which the quantum properties can no longer be understood for inexplicable reasons. Moreover, there is a consensus regarding the assumption that sense-making, clear visual ideas and natural philosophical, reflective thinking are no longer opportune in quantum physics. Quantum mechanics represents a special field dealing with the behavior of smallest particles at smallest scales (at least this was assumed in the 1920s), to which the old ideas no longer apply. Nevertheless, they remain valid as soon as one leaves the level of molecules. Thus the separation of nature came into play.

The actually more obvious idea that the buildings of classical physics and chemistry were erected on foundations which now (experimentally justified) turned out to be *fata morganas* was hardly seriously pursued. Therefore, many physicists actually believe that it is the experimental facts that make a deeper understanding of nature fundamentally impossible, and not unexplained assumptions, faulty model ideas, and outdated physical theories that are supposed to interpret the facts but do not fit with them. This is of course because our four suspects, the atom hypothesis, the division process in wave theory, the theory of gases and the definition of the speed of light, do not appear to the ordinary physicist as suspicious, but as assured findings of physics – quantum theory notwithstanding.

Anyway, what seems to be philosophically so complicated can simply be traced back to the fact that the body conception of mechanics, the atom hypothesis and the traditional wave theory fail because of the true constitution of nature. BOHR's assertion that the human mind had reached a "natural limit of cognition" was thus nothing more than a hasty assumption born of sheer desperation. The real problem was that no one was willing to accept the failure of atomic and particle concepts – and to use Occam's razor accordingly. So, to avoid any confusion in the debate about the physical meaning of quantum theory from the outset, we should always keep the following in mind:

When nature signals to us that there can be no particles and bodies, this does not mean that reality disappears into thin air. It only means that a geometric model – the assumption that EUCLID's geometry describes physical properties – dissolves into air. And when experiments proclaim right before our eyes that atoms cannot exist, it does not mean that matter vanishes into thin air. It only means that the philosophical concept of the atom – the assumption of indivisibility – dissolves into nothing. It is actually very simple: We simply look at the world through the wrong glasses as long as we do not understand that the atomic paradigm is wrong. So what we have to exclude are our false assumptions about the constitution of nature and reality, not the concept of a reality of whatever nature. This is the scientific method.

So if we take the experiments of physics seriously, we will have to get used to the idea that the atom hypothesis has to be thrown overboard, that the body idea of mechanics cannot not really explain the physical properties of matter (not even those of mass), that quantum particles do not exist at all and that traditional wave and field theory is still missing something crucial – our new physical principle of holistic division and branching. This is a new, non-mechanical physical structural property, discovered in experiment, which now allows us to see nature with new eyes. So all we have to do is design a new set of theory glasses based on the holistic division process and compatible with the principle of conservation of energy. Although this seems rather obvious, hardly any physicist could imagine until now that the wiggle picture of quantum mechanics could suddenly flip over and show a new motif. The surprise will be even greater when the disbelieving physicist has to realize that theoretical physics was quite capable of ignoring the experiment, and for trivial reasons: To be able to save the atomic paradigm.

This was and is probably the most persistent scientific prejudice of the last hundred years. What power this paradigm exerts over our thinking, you probably feel already while reading (if you have not already closed the book after the first pages because you consider doubts about the atomic hypothesis as heresy). Apparently, in our modern times we have almost forgotten that creativity, real cognition and scientific progress are always based on the resolution of mental contradictions. The greater the contradictions are, the greater the progress that can be expected. If one follows this epistemological stance, it is relatively easy to make the great discovery that RICHARD FEYNMAN once spoke of – a completely new world view that emerges directly from experiment:

"I think I can safely say that nobody understands quantum mechanics... What we need is fantasy, but fantasy with a terrible straitjacket: if you succeed in discovering another view of the world that agrees with everything that has already been observed, but differs in another respect, then you have made a great discovery. That is almost, but not entirely, impossible." [8]

As we have seen so far, making such a discovery is not as difficult as it seems. The first step is to think about the fundamental ideas we use to try to grasp the constitution of nature. When logical contradictions arise, we have to re-examine the assumptions underlying

8   Richard Feynman, 1965

those ideas. This has nothing to do with mathematics to begin with – that is the conceptual, reflective and creative part of the scientific adventure (called natural philosophy a hundred years ago, or simply *sense giving*). Mathematics only comes into play after that: it has the task of enabling an abstract model construction that can represent causal connections, proportions and physical relationships between certain phenomena and forces, from which invariant ratios can be derived.

The second step is the unconditional acceptance of the divisibility condition in experiment, not only mathematically, but also in the model design with which we attempt to capture the true "physique" of matter and light. This means accepting holistic divisibility as a universal physical principle. This, and not indivisibility, is the fundamental, structure-forming physical property. And this means that we have to throw the *atom hypothesis* overboard. Although the physical disregard of this fact (and thus of the experiment) by the quantum mechanical interpretation seems unforgivable, some hesitation is quite understandable: because if the atomic hypothesis is fundamentally wrong, it must have been wrong already in classical physics and chemistry! But that cannot be, can it? But if it is the way it is – the experiment speaks a clear language – then we just have to explain the empirical findings of classical physics and chemistry without atoms. But how exactly is that supposed to work?

It's clear that the answer must lie in the past. At some point in the history of science, we must have missed something crucial. Consequently, we need to revisit the history of ideas in physics and chemistry to find the appropriate *switches of cognition*. At which point the misinterpretations crept in, we will find out in the next chapter. It is also clear that we then have to explain the molecular structure of matter and the periodic table of elements entirely without atoms. At first glance, this endeavor may seem absurd, pointless or unimaginable to you, especially if you have never heard of molecular orbital theory. Although appearing as a complicated mathematical quantum theory, its message is decidedly clear and simple. It is, by the way, the same message that the double-slit experiment proclaims.

With this insight it becomes clear as well that we have to improve EINSTEIN's provisional quantum hypothesis and add branching and fusion processes in order to understand the true physical structure of light and matter. To do this, of course, one must first know that EINSTEIN's particle-like quantum was only an auxiliary model for the structure of light and served only to model and illustrate local, energetically holistic actions of light radiation. But a hundred years later, this is apparently no longer clear to everyone: a lot of physicists simply skip the deeper thinking and actually believe that light quanta or photons are indivisible "energy particles" flying through space at the speed of light, although even in the quantum mechanical interpretation they are only *virtual* photons (i.e. they exist only in the imagination). However, the fact is that EINSTEIN's structure model was not yet perfect: it could correctly anticipate the energetic wholeness of local actions, but it ignored the division process which temporally and causally *precedes* the absorption event.

Therefore it is simply a myth when some quantum physicists claim that EINSTEIN proved that light consists of particles. EINSTEIN's particle model, i.e. his atomistic interpretation of the energetic wholeness of a point-like acting quantum, is definitely wrong because of the interference condition. So there must be another physical explanation for the quantum-like action of radiation. And this can be – quite obviously – only an energetic symmetry condition. EINSTEIN knew this design and theory problem very well, of course, but could not find the reason for his thinking blockade. Surprisingly, many of EINSTEIN's colleagues never seem to have understood his motivations for searching for the true structure of light and the electromagnetic field in general. It also remains inexplicable for ever how EINSTEIN, BOHR and HEISENBERG alike managed to so stubbornly ignore the divisibility condition of the experiment and outwit their own logical thinking. More on this in the second volume under the section *"The logical tightrope act: How easy it is to fool yourself"*, in which FEYNMAN gives us a deeper insight into this art of thinking.

So, anyone who seriously tries to understand the physical problems of quantum theory should take the following message to heart: One should know EINSTEIN's light quantum hypothesis and his thoughts and comments about it from one's own reading, not just from hearsay. The same applies to AVOGADRO's hypothesis, MAXWELL's theory, special relativity.

If we now apply our branching picture to the history of physics, it also becomes clear that there must have been forks in the path of cognition, where we decided on a certain interpretation of empirical findings and thus on a certain direction of further research. If we then find many decades – or two centuries later – that there is something wrong with these ideas, we can go back in history and see if we can find these forks in the path of cognition where other interpretations were still possible. Once we do, it becomes clear that the cognition and model-design process is iterative, not linear (wisdom long known among designers). And with this we acknowledge also that there is no absolute truth in science and that we are ready to check our ideas and assumptions any times.

And exactly that we will do now. To find out where the forks and switches of cognition are to be found, we have to dive a little deeper into the history of science. There we find valuable ideas that have almost been forgotten and problems that simply could not be solved at that time. We will examine traditional physical conceptions, their origin and meaning, such as the atom hypothesis, the molecule idea, the aether gas concept, the wave concept, the field description, the concept of electric current, and the origins of the definition of the speed of light. We have already discovered the first decisive clue in YOUNG's double-slit experiment, which disproved NEWTON's atomic hypothesis of light. This led to the wave theory whose most important property was divisibility. We will discover the second clue in the founding period of physical chemistry, also two hundred years ago, when chemical reactions disproved DALTON's atomic hypothesis for the first time shortly after its introduction. This is precisely what led to the molecule concept, which had *to embody* the principle of divisibility in the truest sense of the word. The question is only whether molecules are particles at all and whether this molecular divisibility is mechanical or holistic in nature.

Exactly this question is answered by quantum physical experiments, what nobody understands, however. You already guess why: They *prove* that holistic division and branching processes exist in reality, even on the deepest level of nature. So the particles of the quantum theory are only imaginations in our mind. This insight does not exclude that particles can be useful fictions and auxiliary models – at least for a certain time, when one does not know better yet. But this time has already expired a hundred years ago. And this insight also closes the circle to the long lasting debate between Bohr and Einstein about the nature of reality. Now one can understand why Bohr philosophized so mystically and Einstein insisted on a genuine physical understanding of the constitution of light and matter. And therewith it also becomes clear that a *physics without atoms* is quite possible: It is about a structure physics in the spirit of the field descriptions of Faraday, Maxwell and Einstein, supplemented by the new physical principle of holistic divisibility.

So let's take a closer look at the adventure of science. Let us find out where the roots of our misunderstandings lie, how our four suspects are involved in the tricky quantum case and how the shipwreck of the atom hypothesis can already be experimentally justified in classical physics and chemistry.

II. Hidden assumptions and switches of cognition

The Adventure of Cognition: Ideas and Concepts of the 19th Century

When the great American physicist Richard Feynman was asked to think of a single sentence that would convey the most important scientific knowledge we possess, he answered simply: "Everything is made of atoms." Today we take this statement for granted. (Jim Al-Khalili, 2007)

This is exactly what a paradigm looks like: Although the double-slit experiment with single atoms clearly disproves the indivisibility hypothesis, the atom concept is still considered the foundation of our scientific world view. In fact, however, the experiment proves that holistic divisibility is the fundamental principle of nature. This is compatible with molecules, but not with atoms. How this is to be understood is shown by Avogadro's original hypothesis (Mario Wingert, 2011)

Avogadro's molecule hypothesis: Dalton's atoms must divide

If the experiments of quantum physics fundamentally disprove the atomic hypothesis, it must naturally fail in chemistry and biology as well. And indeed, so it is: At the same time that YOUNG experimentally disproved NEWTON's atomistic particle hypothesis of light, chemical reactions of gases disproved the hypothesis of an atomic nature of matter. It took a few years, however, for AVOGADRO to note in 1811 that the atom hypothesis, which DALTON had introduced into chemistry in 1808, failed in these experiments. AVOGADRO replaced DALTON's atom with the integral molecule, which is fundamentally divisible in chemical reactions, and thus created the modern molecule hypothesis. However, the nature of this divisibility could not yet be grasped, because at that time only mechanical divisibility was known. This has remained so until today, in spite of quantum physics, which practically points its finger – the superposition principle – directly at *holistic* divisibility. So stay tuned and pay close attention to how our prime suspect – the atom hypothesis – manages to get off the hook despite overwhelming evidence in quantum physics and chemistry.

As early as the beginning of the 19th century, chemical experiments had shown that substances always reacted with each other in simple, whole-number ratios, which was interpreted as an indication of the composition of matter from whole, in principle countable mass units called either atoms or molecules. Around 1800, and still until 1910, these were interchangeable terms with no clear distinction. The term *molecule* originally meant nothing more than tiny *particles with mass*, i.e. smallest imaginable mass particles of matter. In order to understand the composition of chemical substances and to quantify chemical reactions, it was necessary to design a model of how these particles form various compounds and how substances as water, gases and other materials store heat. A new conception of the nature of heat called *caloricum* had just replaced the ancient idea of a heat substance (phlogiston) and had to be combined now in the framework of Newtonian mechanics with the particle concept.

NEWTON's critics had already pointed out that if mass particles were to exert only an attractive (gravitational) force on each other, sooner or later they would have to clump together. In this model, matter cannot remain in a gaseous state for too long. So, for the particles of gases, a repulsive force had to be postulated to keep the imagined particles at distance. This repulsion was to be produced now by the heat substance called "caloric", which should somehow attach itself to the particle and form a large cushion ball around it, so that the gas particles would be kept apart from each other. In this way, so the idea, one could also explain the storage of heat by matter. In gases, which were assumed to consist of myriads of such particles, the particles, enveloped by their "heat substance sphere", should move in empty space, i.e. in a vacuum. This additional assumption was simply necessary to be able to postulate a mechanical motion of the particles (i.e. change of location), which then allows combinations and recombinations of the particles. So the aim of this *model* was to explain the formation of different compounds with the help of mechanical ideas, i.e. with moving bodies and forces acting between them.

In 1810, JOHN DALTON (1766-1844) first defined chemical elements as substances consisting only of atoms of the same species, i.e. chemical elements are pure, not compounds and cannot be further broken down in other substances. The famous Greek ARCHIMEDES already knew the relationship between volume and weight and was able to calculate the density of materials from this and thus conclude the purity of a substance (in this case, a gold alloy). DALTON, like NEW-

TON a strong proponent of the atomic hypothesis, attempted to use this old knowledge to explain the quantitative relationship of pure substances involved in chemical reactions, with the atomic hypothesis applying to all states of matter, whether solid, liquid, or gaseous. So he assumed that chemical compounds are always formed from *individual* atoms. According to this presupposition, a water particle or molecule should consist of one atom of oxygen and one atom of hydrogen (but actually it contains two parts of hydrogen). DALTON began to rank the known chemical elements according to their mass, which could be determined from density determinations and weighing of the initial and final products of the chemical reaction. The mass of the lightest element, hydrogen (usually a gas), serves as the reference unit with the value 'one' for the relative masses of the elements. The mass of all other elements is thus always expressed as a multiple of the mass of hydrogen. If matter consists of atoms, i.e. indivisible particles, then this mass ratio must also apply to individual atoms. Since mass is conserved in chemical reactions, it should theoretically be possible to determine the number of particles in the starting and reaction products to establish accurate formulas for the chemical reaction. For gases, this raised the question of whether and how the number of particles in a given volume could be determined from the ratio of the masses and volumes of the reacting gases. Only a few years after YOUNG's experiment and only a year after DALTON's definition of the chemical elements, a highly original answer was found that clearly challenged DALTON's atomic concept:

In 1811 the physicist and professor of natural philosophy AMEDEO AVOGADRO (1776-1856) put forward the daring hypothesis that *different gases* of the same volume at the same pressure and temperature always contain the same number of mass particles (divisible or not), which he called *integral* or *constituent molecules.* So he meant mass particles, which at first embody a wholeness. This was a quite surprising and not at all obvious thought, not only because AVOGADRO challenged the atom hypothesis with it, as we will see in a moment: Imagine you swivel a glass jar through oxygen and then close it with a lid. Then you take a second glass jar of the same size and capture another gas, for example hydrogen. Do the same with a third glass jar of the same size and capture water vapor. According to AVOGA-

DRO, the number of particles in all three glasses should be exactly the same! But why should the number of particles of different gases be exactly the same per volume unit? If we were not talking about gases but about sand grains, we would hardly be able to catch exactly the same number of particles twice with the same jar. The probability is almost zero. So what was AVOGADRO thinking?

AVOGADRO's hypothesis was based on the experiments of the chemist GAY-LUSSAC (1778-1850), who had determined the volumes of initial and final products in chemical reactions of certain gases at constant pressure and temperature. For this purpose he used a kind of rubber balloon, in which the newly formed gas could expand at constant pressure. Then, in 1805, together with ALEXANDER VON HUMBOLDT (1769-1859), he made a sensational discovery: Hydrogen and oxygen always combine *exactly* in a volume ratio of 2:1! Two volume units hydrogen combine with one volume unit oxygen to two volume units of gaseous water (water vapor), completely. No hydrogen or oxygen particles or residues remain. This appeared like a miracle of nature, at least if one assumed that gases consist of a huge number of moving particles with irregular varying distances, because that would mean that from the beginning there must have been exactly twice as many hydrogen particles as oxygen particles — as if someone had counted trillions of particles exactly beforehand. In practice, of course, this was completely impossible. So what was going on there?

Clear is that the volume ratios in which the gases react are independent of the size of the unit volume chosen in each case. You can take the unit cubic meter, cubic decimeter (liter), or cubic millimeter, if you work with very small amounts of substances. The volume ratio of chemical reactions does not change by this, which makes it clear that this is a scale-independent *physical property*, a truly universal regularity, which we call invariant behavior or 'natural law'. And this must be true down to the smallest conceivable volume units, which were imagined as tiny bodies and called molecules or atoms. Then, in 1810, GAY-LUSSAC was able to show that many other gases likewise combine in simple integer volume ratios. For instance, one volume part of chlorine and one volume part of hydrogen react to two volume parts of hydrogen chloride (or hydrochloric acid, which is present here in gaseous form). Thus the volume ratio between the reacting elements in this case is 1:1. All this is undoubtedly correct, experimentally well proven and applies still today. And that is the actual brain teaser AVOGADRO has left us: According to Avogadro's hypothesis, one liter of hydrogen chloride (HCL) should contain the same number of particles as one liter of hydrogen (H) or one liter of chlorine (CL), if temperature and pressure are kept constant. And since their chemical reaction makes two liters of hydrogen chloride, the volume of the reaction product must now contain twice as many chlorine and hydrogen particles: 1 H + 1 CL makes not 1 x HCL, but 2 x HCL, thus producing 2x H and 2x CL. So far, so good. Sounds simple – so what's the problem?

Well, the punch line of this joke is easily overlooked these days, but contains the deepest secret of nature: That undoubtedly means that the chlorine and hydrogen particles – DALTON's hypothetical atoms – must have *doubled* during the chemical reaction. One becomes two! How is this possible, if DALTON's mass particles are supposed to be *indivisible*? That's exactly what AVOGADRO was wondering... And since no real natural scientist believed in miracles yet at that time, AVOGADRO did not speak of "virtual atoms" either, but concluded that DALTON's *atoms must have divided during the chemical reaction*. This clearly means that the hypothetical mass particles or molecules cannot be atoms!

If you haven't quite got the punch line yet, just imagine – like DAL-TON – that the initial volumes each contain only *one* chlorine atom (CL) and only *one* hydrogen atom (H). The chemical reaction produces two compound molecules hydrogen chloride (2 x HCL), one per volume unit. Since each HCL compound molecule consists of one chlorine and one hydrogen atom, but two of them were formed, the number of chlorine and hydrogen atoms must have somehow doubled during the chemical reaction. How is that possible? What explanation would you have proposed if you had been a natural scientist at the time? Is nature showing us her true face here? I would say yes – don't you see it? (Fig. 8). Obviously, chemical experiments disprove DALTON's atomic hypothesis, unequivocally.

That was the discovery that Avogadro had made. It was the birth of our modern molecule hypothesis, whose most important physical property is divisibility: molecules are volume units with a certain mass that normally embody wholeness but can – and must – divide in chemical reactions. And this raises the crucial question: *What is the true nature of this division process?* Is it a mechanical division, a splitting, or can we explain the doubling of particles in another way? We would certainly not fall for the idea of calling them *virtual particles*, as in quantum mechanics... So how to explain this strange doubling process? How and why can a whole suddenly divide, but only during a chemical reaction (or at a double slit)? What kind of division is this? This is the question that Avogadro did not find easy to answer. In order to be able to explain – or at least illustrate – the molecular division that undoubtedly takes place, he put forward a second hypothesis, which, however, could not be derived from the experiments, as we will see in a moment.

What we can recognize today in chemical reactions, if we look at them with 'quantum eyes', is mainly this: There, too, the question is whether the division of the mass particle is mechanical or holistic in nature. Interference and diffraction experiments of quantum physics (and their wave equations) show that atoms and molecules must be divisible in a non-mechanical, holistic way. We find it difficult to believe the testimony of experiment because it contradicts the atom or indivisibility hypothesis, which we hold in higher esteem. The atomic hypothesis, in turn (now meaning a mass particle), seems to be supported *nowadays* by Avogadro's molecule division, which is clearly proven by chemical reactions. Thus, the logical chain of conclusions seems to be watertight, leading to confusion, perplexity and sheer despair in the interpretation of quantum physics. But the solution of the Gordian knot is quite simple. Somewhere in the logical chain unproven assumptions must be hidden. And that is exactly how it is, as we will see in a moment. Avogadro's molecule hypothesis can indeed be interpreted holistically if it is freed from ad hoc assumptions that were neither experimentally justifiable at the time, nor can be experimentally proven today, as quantum double-slit experiments with single atoms show. This insight leads straight to the discovery of molecular cell division (Fig. 9).

Figure 8: Avogadro's brain teaser: What's wrong with Dalton's atoms?
The chemical reaction of chlorine and hydrogen gas shows that one volume unit of chlorine and one volume unit of hydrogen react to two volume units of hydrogen chloride gas. If gases consist of atoms, as Dalton assumed, how can the doubling of the atoms in the reaction product be reconciled with the indivisibility hypothesis? If one imagines only one atom per volume unit as in this illustration, the problem becomes particularly clear: Dalton's atoms must have doubled by division during the reaction!

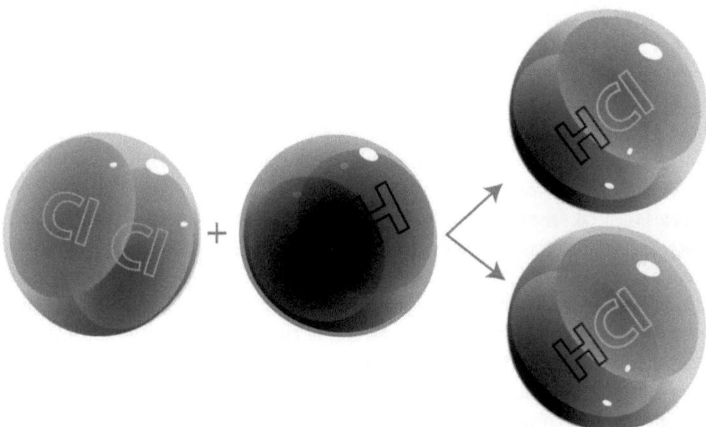

Figure 9: Avogadro's molecule hypothesis, 1811: Chlorine and hydrogen atoms that divide
Gases with the same volume, pressure and temperature always contain the same number of whole molecules, also called integral or constituent molecules. Although molecules are normally stable (like atoms), they are potentially divisible in chemical reactions. Avogadro could also have said that Dalton's atoms are divisible, but since atom means 'the indivisible', he avoided this term completely and spoke of divisibility of integral (whole) molecules. The crucial question is whether we have to interpret this divisibility as a mechanical dicision of an already preconfigured (bipartite) molecule or as a holistic division of a homogeneous, non-preconfigured integral molecule. If the latter is true, we have discovered molecular cell division.

Looking over Avogadro's shoulder

Perhaps I should mention here that it was a very long time ago that I first suspected that the atomic hypothesis was wrong (1978). Of course, I knew also the traditional scientific views: During that time I attended the 12th grade of the Gymnasium (also called Extended Secondary School), completed a thorough scientific basic education, and graduated with good Abitur grades. These included chemistry, physics, astronomy, biology, mathematics etc., including the basics of molecular orbital theory and matrix calculus. In parallel, I took optional courses in solid states physics and biology. However, it was not until eight years later that I accidentally stumbled upon the trail that led directly to AVOGADRO's original theory and shed new light on the failure of the atom hypothesis in quantum physics. Apparently, the old wisdom still holds true: Seek and you shall find.[9]

9 In the 11th and 12th grade I was given the opportunity to take a 'quantum course' at the Institute of Solid State Physics of the Academy of Sciences in Halle/Saale, Germany (today a Max-Planck-Institute), since I still wanted to study physics at that time. Topic: "Development of a computer program to analyze electron diffraction images for crystal drawings." The most valuable treasures of this institute were an electron microscope in the basement (probably by Manfred von Ardenne), with which electron diffraction experiments and crystal analyses could be carried out, and desktop computer in front of which physicists queued up. This was in 1978/79 (the 'computer' was a typewriter-sized Hewlett-Packard desktop calculator with only four registers). In this context I naturally thought about the wave-particle paradox and realized that the atomic world view had somehow to be wrong in general. I also realized that solving the quantum puzzle was no longer on the agenda in physics, however. That was the moment when I, at the age of just 17, decided not to study physics. I had just passed the artistic aptitude test at the University of Art and Design Burg Giebichenstein and had an admission to study industrial design in my pocket. From time to time, however, I continued to search for clues as to how my still vague idea of branching processes at the double-slit could be reconciled with the ideas of chemistry. Could the atom hypothesis somehow be shaken by chemical experiments? Then, in 1986, the following line jumped out at me: **"According to AVOGADRO, N particles chlorine and N particles hydrogen produce 2N particles hydrogen chloride. Consequently, the particles of chlorine and hydrogen must have divided during the reaction."** (Struktur der Materie. Kleine Enzyklopädie. VEB Bibliographisches Institut Leipzig 1982; a book which I bought myself as a Christmas present in December 1986). I was completely stunned when I stumbled upon this sentence, although I knew this fact from chemistry, of course. It was the particular wording that referred to the genesis of molecular theory: mass particles (Dalton's atoms) that divide... That was the lead of the prime suspect I had been looking for, and the first concrete clue that confirmed my suspicion that we are missing something crucial in the history of physics and chemistry: The failure of the atomic hypothesis and the existence of holistic division processes, as they occur everywhere in nature – and could hardly be overlooked even in electron diffraction experiments, which I had dealt with some years before.

To find out how AVOGADRO explains the molecular division, how he deals with the atom hypothesis, and whether he was already thinking of holistic division processes, I tried to track down the original essay – perhaps this paper could answer the question why the idea that the atomic hypothesis could be wrong has never been seriously pursued. I finally discovered it in an anthology of the German chemist WILHELM OSTWALD (1853-1932) from 1902, entitled "Die Grundlagen der Atomtheorie".[10]

The first time AVOGADRO's essay is difficult to read and not easy to understand. On the one hand, the new terms were not yet exactly clear, on the other hand the writing style is ambiguous and somewhat difficult to get used to. AVOGADRO himself doesn't yet seem to know exactly what all this means: You can literally look over his shoulder as he gropingly searches for suitable pictorial ideas for this strange division process of matter he had discovered. It was the early days of physical chemistry: There was no clear distinction between atoms and molecules, many elements had not yet been discovered, and the periodic table of chemical elements did not yet exist. AVOGADRO's molecule concept sheds new light on the mysterious divisibility of matter, which was to catch up with us again only one hundred and fifteen years later in quantum physics.

There, the completely unexpected behavior of matter left physicists dumbfounded and speechless until this day: atoms and elementary particles like electrons seemed to be clearly divisible, but in a weird and mysterious way that seemed to make no physical sense. That's the wave-particle paradox, the contradiction in our ideas between the divisibility of gases and fields and the indivisibility of imagined atomic particles. However, in order to see this connection between molecular and quantum theory, one must know AVOGADRO's original hypothesis. AVOGADRO shows how, despite many unclear details about the true constitution of matter, important physical principles can be drawn from the experiment; in this case, that the constitu-

10 Wilhelm Ostwald: "Foundations of Atomic Theory". Today, Avogadro's original hypothesis can easily be found in the internet: A. Avogadro: "Essay on a Manner of Determining the Relative Masses of the Elementary Molecules of Bodies, and the Proportions in Which They Enter into These Compounds." (web.lemoyne.edu/~giunta/ea/AVOGADROann.HTML).

ent volume units with mass, aka mass particles, must be divisible, i.e. cannot be of atomic nature, and that this divisibility must be a fundamental property of matter. But how did he come up with it, exactly?

GAY-LUSSAC's experiments had revealed the surprising property that gases always react with each other in very simple volume ratios of 1:1, 1:2 or 2:1. As already mentioned, oxygen and hydrogen react in a volume ratio of 1:2, chlorine and hydrogen in a ratio of 1:1. These volume ratios of the reacting gases, which could be grasped in simple integers, now made it clear to AVOGADRO that this wonder of nature could not depend on tiny details of the constitution of the gases – for example, variable distances between particles, a random number of particles, or the presence of a hypothetical heat substance that would envelop and separate particles. On the contrary, leaving aside these speculations, which cannot be clarified experimentally, there seems to be only one possible way to associate the surprisingly simple volume ratios with mass and density determinations and the conservation of mass: There had to be some principle behind it.

AVOGADRO captures this principle with the bold assumption that the number of *integral molecules* (undivided volume elements with mass) does not depend on the type of gaseous matter and is therefore the same for all gases at the same volume, pressure and temperature:

"The hypothesis which is presented here at first sight and which even seems to be the only admissible one is the assumption that the number of whole molecules in each gas with the same volume is always the same, or always proportional to the volume."

This means practically that in all other, more complicated explanatory attempts, the simple volume ratios of reacting gases would still appear to be a wonder of nature. Basically, AVOGADRO has thus used an invariance or symmetry principle, like GAY-LUSSAC in the experiment: The condition of the same pressure and temperature ensures that the densities have not changed and therefore are comparable before and after the chemical reaction. This also makes it possible to segment the volumes (in mind and experimentally) into any number of small volume sub-units of equal size with a corresponding mass. If one imagines these volume segments very tiny, one may consider such a volume unit as that space that one mass particle or molecule occupies (as in figures 8 and 9, where the volume is spherical and identical to the mass particle, atom, or molecule).

Interesting is now that this simple principle functions completely independently of the question of whether the assumption about the existence of tiny particles was correct or necessary at all. And this has practical consequences:

"Assuming this hypothesis, one can very easily determine the relative masses of molecules that can be obtained in gaseous form, as well as the relative number of molecules in the compounds. Because the ratios of the molecular masses correspond to those of the densities of the gases at the same pressure and temperature, and the relative number of molecules of a compound is directly given by the ratio of the volumes of the reacting gases."

Exactly this leads to the conclusion that the atoms of DALTON must have divided during the chemical reaction. But how should one justify this divisibility? To this end, AVOGADRO put forward his second hypothesis, which is completely independent of the fact of divisibility, but *cannot be deduced from experiment*: If the integral molecules were already composed of smaller components, the division could be explained mechanically as decay into these parts:

"A means of explaining facts of this type in conformity with our hypothesis presents itself naturally enough; we suppose, namely, that the constituent molecules [...] are not formed of a solitary elementary molecule [meaning Dalton's atom], *but are made up of a certain number of these molecules united by attraction to form a single one, and [...] divide itself into two or more parts..."*

So AVOGADRO's most important discovery was that during chemical reactions a division process of these 'atoms' must take place. As he expressed it: *"A division of molecules of which this physicist* [DALTON] *had no idea"*. Only then can the volume equations of GAY-LUSSAC be explained, which were simply not compatible with the indivisibility assumption of (still hypothetical) particles or volume elements with mass. This shows that the atom hypothesis already failed in the early days of chemistry, and that AVOGADRO deliberately did not speak of atoms, but of mass particles that are *divisible in principle*, which he consistently called molecules. Thus, the term molecule becomes a synonym for divisible particles or volume units with mass and can be clearly distinguished from the term 'atom' since then. In this way AVOGADRO designs a *chemistry without atoms* that only operates with molecules. AVOGADRO distinguishes three types:

1) *Integral molecules*
   These are gas molecules in a stable, undivided state that operate as a whole. Also referred to as the *constituent molecules* of a gas.
2) *Elementary molecules*
   These are the division products of integral molecules which then have the smallest *relative* mass (our atoms today).
3) *Integrated molecules*
   These are the compound molecules of the reaction products, i.e., composite or fused molecules.

The use of these terms appears ambiguous and somewhat confused when reading the original essay for the first time, but can then be understood quite clearly in this way: Constituent molecules are the units of which the gases are made up, i.e. molecules in the sense of today's chemistry. In the normal state, they behave like stable, holistically operating units – just like atoms. AVOGADRO therefore calls them also integral molecules. Yet these molecules can divide during chemical reactions, so they have the *potential ability* to divide. This is also reflected in our modern scientific view, but is now interpreted as the mechanical divisibility of diatomic molecules. In AVOGADRO's view, the term atom plays no role: The division process of an integral molecule produces elementary molecules, which are the smallest determinable relative masses of the elements in the compound (which are equivalent to modern atoms). The fusion of different elementary molecules or of elementary molecules with other constituent molecules forms integrated molecules, i.e. composite molecules or compounds in our modern sense. All molecules that form a gas volume can be seen as constituent and integral molecules, whether they are pure chemical elements or chemical compounds. It is the number of these undivided (integral, holistic) molecules that should be the same in all gases at the same volume and temperature.

In reading AVOGADRO's work, one could easily overlook the fact that the presupposition that gases are made of discrete particles remains an unconfirmed conjecture despite his discovery of division processes. There was no experimental proof of the existence of particles, let alone of atoms or molecules. It has only been proven that different gases exist with certain densities and chemical properties, that their volume units can divide in chemical reactions, and that this divisibility property does not depend on the size of the chosen volume unit, what excludes the atomic hypothesis with certainty: Even the smallest conceivable volume unit is divisible. Thus it becomes clear that these experiments are basically only about volume units with mass, which we could also call 'density blocks' or 'volume weights'.

This is true even if the mass were concentrated in the center of the chosen volume subunit and surrounded by a vacuum, as the particle picture suggests, because the vacuum weighs nothing. In any case, the experiments do not make any statements about whether these molecules, these volume elements with mass, exist in a bodily form at all, i.e. whether the Newtonian model of tiny billiard balls, divisible or not, applies to reality at all. And they say nothing about how big a single molecule really is or how many molecules are contained in a certain volume. So the molecule, which we imagine as bodily object, is still not more than a spawn of the thinking mind.

"Wait a minute, you may object here, I have learned that Avogadro's hypothesis is a law, that Avogadro's molecules are made of atoms, and that Avogadro can quantify the number of particles very precisely".

Yes, of course, that's what I learned too (except for the latter; that's LOSCHMIDT's number). Of course, what we learn can sometimes be wrong, or needs to be updated from time to time, otherwise there would be no scientific progress at all. We will come to the atomic interpretation in a moment, and to the question what this number is all about. In any case, from the original text we can see that Avo-

GADRO's hypothesis that equal volumes of different gases contain the same number of sub-units is not a law, but simply an application of the relation between volume and weight, i.e. the concept of density. That application made possible the very hypothesis that these volume units must be divisible in two molecular parts in reactions, which explains why they react with each other in very simple integer volume proportions. This was a real discovery: the fundamental divisibility of gaseous matter in chemical reactions.

Why Avogadro cannot determine the number of molecules

So what AVOGADRO really has done is this: One can mentally divide a gas volume into any number of smaller, but always equally sized volume units, which all have the same mass or weight. Equal volume units of different gaseous chemical elements or compounds differ in weight. That sounds almost banal, because that is the definition of *density*. This physical property, defined as mass per volume unit, is independent of scale, so the density remains the same in any chosen volume. For instance, the density of water is 1 gram per cubic centimeter, 1 kilogram per liter, and 1 ton per cubic meter (this is no wonder of nature, but a definition, since the density of liquid water is chosen as the unit of comparison). This fact can be used to express the *relative density* or specific mass of substances, which is therefore a number without dimensions. For water, the density is always 1, no matter which volume unit is chosen. To make such a comparison: The relative density of medium-weight wood as pine or linde (basswood) is about 0.5, of aluminum 2.7, of iron 7.9 and of gold 19.3. If you now imagine a cube of one meter side length, you can immediately say that such a cube of wood will weigh 500 kg, of water 1,000 kg, of aluminum 2,700 kg, of iron 7,900 kg, and of gold 19,300 kg. But if it's only a cube of ten cm side length (which is a liter), then its weight is 0.5 kg / 1 kg / 2.7 kg / 7.9 kg or 19.3 kg, respectively.[11] For instance, if someone asks you what a square meter of gold sheet with 1 mm thickness weighs, you can answer in a flash: 19.3 kg (19,300 kg / 1,000 sheets à 1 mm) and know at the same moment that it has a volume of one liter.

11 The specific mass enables designers who have the numbers of usual material densities in mind to have handy pictorial conceptions of material properties and to make quick rough calculations.

So if you have a gas with a certain density and volume, you can arbitrarily decide into how many volume subunits with mass or '*density blocks*' you want to mentally – or experimentally – divide it. Do you want to divide a liter into 1,000 cubic centimeters, or rather into 1,000,000 cubic millimeters? If you had an idea what volume a single molecule occupies and how many there are, you could subdivide the volume appropriately. But that doesn't matter at all if you want to determine the *relative densities* and therewith the *relative masses* of the volume units with mass that react with each other, because the density always remains the same, regardless of the chosen subdivision scale. So how large the number of molecules – or volume subunits with mass – per unit volume really is, cannot be determined for this reason: This number should only be equal for all gases. This is not a weakness of Avogadro's hypothesis, but only an expression of the fact that the density at the same pressure and temperature is independent of the chosen spatial (volume) scale. It follows that gas volumes of any size must also be 'molecularly' divisible. Thus one can assume any number of subunits of the total volume, but also use the volume unit one, as in Fig. 8/9, without to specify their size, to be able to represent the sum formulas of chemical reactions in *very simple* integers – after all, we are only dealing with the numbers 1 and 2! In these illustrations, the unit volume is a sphere which is identical to the "atom" or integral molecule. This is obviously not about higher mathematics, but about *holistic bipartition, bifurcation or branching processes*. So the real reason why the volume ratios are so simple and always integer is that *any volume of a gas* can divide into two 'molecular halves' in chemical reactions. Therefore, with Avogadro's hypothesis, no statement can be made how big a molecule 'really' is, or how fine the molecular structure of a gas might be. And that is the punch line:

Since this picture is scale-less, the holistic molecular division principle would also apply if these gases were originally homogeneous, i.e. without any internal structure, just like the atom in fig. 8. The next bipartition cycle would then act on two subunits simultaneously, what makes four subunits. And the next would make eight subunits, and so on... This resembles a *biological* cell division and fractal properties. Thus, in my view, the most important thing is the experimentally required physical property of the divisibility of gaseous matter, which creates a molecular (i.e. bi-partitioned physical) structure in the first place. We leave aside for the moment the fact that temperature must also produce structural changes, as well as the fact that a holistic molecular division can occur even if the molecule has already been divided once or if the integral molecule has an inhomogeneous internal structure (like a biological cell). This is the core of AVOGADRO's concept, which leads to idea of chemical reactions taking place in the form of *division and fusion processes* of volume units with mass, called molecules. In the course of such considerations, one also realizes that the question of the nature of atoms and molecules is not at all about the smallest things in nature, but about scale-less structural properties that might apply to arbitrary size ratios.

AVOGADRO's molecule division concept made it possible for the first time to determine the exact ratio of elements in a compound, which he also demonstrated for water, ammonia and other substances. He shows how the relative masses of molecules can be calculated from the density and volume ratios of the gases which unite in chemical reactions. He further shows that, in comparison to DALTON's calculations, some molecular weights have to be doubled (because of the division). DALTON relates the mass determination to the mass of the hypothetical hydrogen atom as unit. But since this atom had to be divisible, the molecule mass had to be twice as large if the density and volume ratios are taken into account, and the principle of conservation of mass applies to chemical reactions. Thus, the mass of DALTON's hydrogen atom, which AVOGADRO had recognized as divisible and named integral molecule, had to be doubled. Since an *elementary molecule* is the product of this division, it weighs only half (this is our present hydrogen atom). So AVOGADRO relates his mass determinations to the divisible hydrogen integral molecule as unit, which we simply call hydrogen molecule and mark with the symbol H_2.

However, since the masses can only be determined as ratios and the true number of molecules in the gas volume is not known, AVOGADRO would not only allow sum formulas of water such as H_2O, but also H_4O_2 or H_8O_4. The indices could also be larger integers, only the proportions $(2:1)$ must remain the same. How big and heavy a constituent molecule really is, or how many integral molecules the volume unit should have consisted of originally, can therefore not be said with AVOGADRO's hypothesis. However, both AVOGADRO and the modern chemist express the sum formula it that way: Two volume units of divisible hydrogen $(2\,H_2)$ react with one volume unit of divisible oxygen $(1\,O_2)$ to form two volume units of water vapor $(2\,H_2 + O_2 = 2\,H_2O)$. This is the empirical part of GAY-LUSSAC's volume equation, which incorporates AVOGADRO's finding that molecules are divisible.

All this shows furthermore that the little index sign of the symbol H_2 can carry two different physical meanings, depending on how one interprets the molecule division: Either it symbolizes a molecule which already consists of two components, as in AVOGADRO's ad hoc hypothesis and in the atomic world view, or it symbolizes the possible divisibility of a homogeneous molecule *into* at least two main volume parts, which leads to a non-atomistic, holistic world view. In the latter case, we should reintroduce AVOGADRO's notion *integral molecule*, since it can replace the term atom without any problems. We will discuss this amazing possibility in more detail in a moment, after we have learned how the atomistic interpretation of the molecule division process came about in the first place.

So at this stage, we have still two options to interpret the molecular division: Either it is a mechanical splitting, a decay in two or more separate parts, or it is a holistic division process, as known from cell division and branching in biology, where the parts remain a physically coherent whole. This is at least my proposal, inspired by the unsolved problems of quantum theory. For chemistry and physics, this is a novel hypothesis that may seem strange, if not crazy, at first glance. What is clear for both variants, however, is that the molecule embodies a physical-chemical entity that operates as a whole and divides only in chemical reactions. To explain the division, AVOGADRO made his second assumption, which states that integral molecules consist of smaller molecules somehow connected by attraction, which divide during the chemical reaction into two or more *elemen-*

tary molecules. However, if we follow the idea of a holistic division they should rather be called *secondary* or *partial* molecules, just as we speak of secondary or partial rays when a light beam divides, but they could also be called daughter molecular cells if we think on cell division in biology. So AVOGADRO's additional assumption was not as "natural" as he had believed, because nothing can be concluded from the chemical experiments about the structure of the integral molecules *before division*. The assumption of a preconfiguration of the integral molecule was thus only an ad hoc hypothesis, a shot in the blue, to be able to illustrate the division mechanically. And this makes it clear that a holistic molecule division is equally compatible with GAY-LUSSAC's chemical experiments and AVOGADRO's division hypothesis.

Reading AVOGADRO's paper, one can get the impression that he was already thinking vaguely in the direction of a natural division process. He had already imagined how a whole molecule divides into halves, quarters, eighths and so on, as if a whole organically divides here. However, it is not clear whether this molecular division occurs all at once, or whether it proceeds step by step as in biological cell division (a mother cell divides into 2-4-8-16-32-64- etc. daughter cells). At any rate, this conception would fit in with the indeterminate sum formulas of water such as H_4O_2, H_8O_4 etc., and with his idea that the resulting water steam compound molecule must divide itself again, for instance H_4O_2 in 2 H_2O to make two volume parts. He had also suspected that the divisibility of such 'macromolecules' could be a natural principle, which serves to avoid larger mass concentrations. However, this could also be interpreted as molecular structure formation, in which compound masses reorganize themselves optimally at the lowest possible energetic state.

What AVOGADRO could not have known then and we realize only now, two hundred years later: More likely, and equally compatible with experiment, is the hypothesis, now supported by our quantum physical insights, that the division of the integral molecule is a holistic division process, no different in principle from cell division and branching processes in biology. In 1811, there were still no alternatives to the ideas of mechanics, because biological cell division was not discovered until around 1838. Decades passed before it was fully accepted scientifically. Fifty years and much better microscopes were needed to at least partially understand biological cell division. Incidentally, the first movie that could actually document *the process* of cell division was not made until 1943. And the giant macromolecules of DNA (genes), which divide lengthwise in the cell nucleus and thus initiate cell division, were not discovered until 1953. Thus it seems that AVOGADRO was right with his visionary idea of macroscopic molecular division, even if it does not apply to all macromolecules, let alone water (at least as far as we know today). But all this happened in a branch of the natural sciences to which physics and chemistry could not yet see a direct connection.

Summary:

AVOGADRO discovered that gases must be divisible in simple integer ratios in chemical reactions, regardless of the size of the chosen volume units. If gases consist of tiny particles or volume elements with mass, this must also be true for the constituent particles or volume elements. So it turned out that DALTON's hypothetical atoms were divisible. AVOGADRO therefore called them not atoms but integral molecules, which are stable in the gas state but potentially divisible. To illustrate this physical property of divisibility, he assumed that integral molecules are preconfigured, i.e. composed of elementary molecules, but with no experimental evidence for this. AVOGADRO's elementary molecules correspond to our modern atoms, which are supposed to exist invariably since eternity, but in AVOGADRO's concept they are only the division products of integral molecules.

That gases should consist of tiny, spatially separated particles moving in a vacuum was a fundamental premise of Newtonian physics that was never questioned by AVOGADRO or anyone else. This presupposition is still today the basis of the kinetic theory of gases and the mechanical theory of heat. Since the phrase *divisible atom* is a contradiction in itself, AVOGADRO never used this notion, although this could have helped to understand his reasoning. Remember that a wit consists of a sharpened contradiction that the mind can only resolve through a new conception. This is what we call insight or cognition: The experimental realization that atoms must be divisible leads in the physical theory to a logical contradiction with the paradigmatic assumption of indivisibility. At this contradiction we will despair eternally, if we do not recognize our atomistic prejudice as such. However, if we replace the word atom by the term integral molecule, it becomes immediately clear that the empirical content remains the same: In the gaseous state, we are dealing with stable, physically integral units which, however, are chemically divisible.

It remains correct that AVOGADRO's molecular division concept can be explained either mechanically or holistically – both views are compatible with chemical experiment. However, the experimental findings of quantum physics made between 1991 and 2003 (double-slit and interference experiments with single atoms and massive macromolecules) now finally force us to consider the holistic divisibility of atoms and molecules as confirmed, thus proving both to be integral molecules in AVOGADRO's sense. However, just as in biology, it must be a matter of holistic division processes *en gros*, which gives a new perspective (e.g. if a cell cluster of 128 cells divides, 256 cells are created). This means that holistic divisions, since they are independent of the number of already existing structural elements, must be scale-less natural processes. We are therefore dealing with a universal bifurcation phenomenon.

How Cannizzaro saved the atomic hypothesis

AVOGADRO understood integral molecules as particles and volume elements with mass that are potentially divisible and reconstructed this physical property from experiment. Proceeding in this way, one does not necessarily arrive at the atomic hypothesis, at a philosophical assumption that deliberately excludes divisibility, but at a conception that includes divisibility on principle. This is a path of cognition that science has completely overlooked. AVOGADRO himself did not evaluate the consequences of his hypothesis for the atomic world view, even later nobody did it. But his hypothesis raises an absolutely serious question as early as 1811, based on experimentally secured facts: are atoms divisible? Are they atoms at all? It was probably this question that every reader asks himself when studying

AVOGADRO's original essay, that has led to his hypothesis being completely ignored for fifty years. To believing atomists, the divisibility hypothesis seemed crude and absurd, if not foolish.

But would the reaction today be so different? Many, if not all physicists still believe in an atomic world view despite double-slit experiments with individual atoms and all the other unsolved problems of physics. However, science is not about belief, but about temporary hypotheses and experimental evidence. If hypotheses are regarded as too certain, as completely secure for all time, then they can certainly lead to a quasi-religious belief or ideology. It is precisely this belief in the atomic nature of reality that prevents us from solving the quantum puzzle that still poses us – now at the double-slit – the same question: Are atoms divisible? Are particles real at all?

To this day, historians are stunned to wonder why chemists ignored AVOGADRO's ideas for so long. The problem was that NEWTON's particle model could not apply to reality, as his critics had noted early on. If there were only attractive forces between the bodies, the universe – like every gas volume – would have to collapse rather fast. Therefore, someone had introduced the repulsive force between the gas particles to keep the particles at a distance. When AVOGADRO claimed that gas particles could somehow stick to each other to form normally stable molecules, this contradicted the distance hypothesis and the assumption of a caloric heat cushion. The same problem, by the way, still exists: In EINSTEIN's general theory of relativity, the repulsive force is called only *cosmological constant* and has remained the greatest mystery to this day – not of chemistry, mind you, but of cosmology! AVOGADRO did not live long enough to witness the success of his molecule concept, although ANDRE-MARIE AMPERE (1775-1836) argued similarly in 1814. Only after half a century of silence did chemists begin to accept the idea of molecular division. This was the merit of STANISLAO CANNIZZARO (1826-1910), who presented, explained and re-interpreted the molecule hypothesis in 1860 at the first International Congress of Chemistry in Karlsruhe (Germany). CANNIZZARO's success was due to the fact that he had succeeded in reconciling the strange divisibility of molecules with the atom hypothesis (Fig. 10):

Figure 10: The atomistic-mechanistic molecule hypothesis, 1860 (Double atoms, Cannizzaro)
Fifty years later, Cannizzaro combined Avogadro's hypothesis with the atom hypothesis: Dalton's atoms should not divide, but would already exist twice from the beginning. This is the atomistic-mechanistic molecule theory as we know it today. The two atoms now seem to be connected either mechanically or by strange binding and attraction forces.

DALTON's atoms would not divide (this seemed completely absurd) but should exist doubly from the beginning. The integral molecule should consist of two atoms that behave normally like a whole unit but can divide in chemical reactions. CANNIZZARO apparently took AVOGADRO's *ad hoc assumption* of a preconfiguration of the molecule seriously (still without experimental evidence) and interpreted AVO-GADRO's hypothetical elementary molecules, which were originally the division products of an integral molecule, as the correct atoms. Pretty clever! Practically he confirmed the failure of DALTON's original atom hypothesis, but reintroduced it in the same breath on a freshly invented *deeper level of nature*. And thus the atomic hypothesis was saved!

In this way, AVOGADRO's original hypothesis, which clearly contradicted the indivisibility assumption, could be erased from memory without having to change anything about the quantitative relationships he had derived from it. CANNIZZARO's reinterpretation of the integral molecules of gaseous chemical elements as diatomic molecules implied as yet unknown attractive or bonding forces between these atoms, but also a mechanical divisibility that leads to separate, individual atoms. This fit perfectly into the world view of Newtonian mechanics, in which almost all scientists still believed at that time.

CANNIZZARO's concept, though philosophically and logically tricky, was a real liberation blow that quickly led to great progress. Chemical reactions could now be better understood, modeled and visualized. CANNIZZARO's merit lies above all in the fact that he declared AVOGADRO's postulate of the same number of particles in different gases to be an indispensable heuristic principle and the division of molecules in chemical reactions to be an experimentally established fact. This enabled a better understanding and methodical treatment of chemical reactions. So progress is possible if the underlying model conceptions are useful and consistent with the experiment, even if they are not entirely correct and maybe vague, or if they turn out to be wrong one day. That is the very nature of hypotheses. With the recognition of AVOGADRO's molecular division concept and the atomistic interpretation by CANNIZZARO, chemistry experienced a tremendous upswing that would soon lead to new industries, especially in Germany. Just thirty years later, around 1890, it dominated world trade in synthetic dyes and chemical products of all kinds.

Since then, the indivisibility hypothesis seems to be compatible with the divisibility hypothesis, at least in the framework of Newtonian mechanics. This is the atomistic-mechanistic model of the molecular structure, which still seems to be valid – but only if one disregards the molecule orbital theory and interference experiments. The atomistic interpretation is an sound explanation and still the official one, but not the only possible one. Despite all the theoretical and practical successes, many physicists and chemists at that time were still clearly aware that no one had ever seen an atom or a molecule. It was practically agreed that these conceptions, while undoubtedly useful models, did not necessarily represent the true constitution of matter. In contrast, many physicists and chemists today feel far too confident with regard to the validity of the indivisibility hypothesis and the atomic world view. This development began around 1898 with the invention of the electron particle and is also related to the work of EINSTEIN and PERRIN. It was primarily their work that led to the full recognition of the molecular nature of gaseous and liquid matter *in physics* around 1910 – and thus indirectly the existence of atoms as well.

Originally, this meant nothing else than the confirmation of the assumption that gases and liquids are not homogeneous but have a certain structure. This structure, for instance that of water, appears to be in permanent internal motion at room temperature, though the surface is smooth as a mirror. This phenomenon became known as BROWNIAN motion. At the beginning of the 20th century, this motion could be made visible under the ultra-microscope, in which the movements of tiny bodies suspended in water, such as plant pollen, dust corns or mastic balls (with a diameter of only 1/1000 mm), was followed. According to the theories of EINSTEIN and PERRIN, these tiny corpuscles are randomly pushed around due to collisions with trillions of water molecules. This was interpreted as a definite confirmation of the molecular structure of matter *and* the kinetic theory of gases and liquids. Since molecules were supposed to consist of individul atoms, the existence of atoms was accepted as well, at least indirectly. Thus, the existence of corpuscular molecules and indivisible atoms seemed to be confirmed by physical phenomena as well, quite indepent of chemistry.

At this point, however, it was still clear that they are constructive theories. They are designed on the basis of certain assumptions that serve to model the phenomenon in question without contradiction. In this case, the theory of Brownian motion shows that the assumption of moving particles – i.e. the motion theory of mechanics – and the molecular particle hypothesis are compatible with experiment. The truth of the theory thus acquires a high probability, as long as no new facts become known that contradict it. Nevertheless, this is by no means a direct empirical proof for the existence of corpuscular molecules, let alone of indivisible atoms, since these cannot be seen even under an ultramicroscope. Therefore, the usefulness of such theories could not convince all physicists that the atom hypothesis is physically correct. Others saw in the atomic assumption a sufficiently secure basis on which to proceed to fathom the nature of matter, although the wave theory pointed in exactly the opposite direction. This led soon to deep contradictions.

Surprisingly, all the findings of quantum theory between 1905 and 1927 did not change the widespread conviction that matter on the deepest level consists of tiny indivisible corpuscles, even though all the alarm lights were frantically flashing red. This inability to challenge the paradigmatic atomic assumption, despite clear evidence, led to our current understanding of science: since then, it has been assumed that the findings of quantum physics do not overturn the atomic world view, but only modify it, and only at a 'submicroscopic level of nature' introduced especially for this purpose. This interpretation, quite questionable from the point of view of natural philosophy, overlooks the fact that Avogadro's original idea of divisibility of integral molecules,, Young's double-slit experiment, and quantum theory itself call into question the very ideas and assumptions that underly the atomic-mechanistic world view, namely the existence of *discrete* particles and their *indivisibility*.

Now, if you try to discuss Avogadro's hypothesis and the failure of the atomic assumption with a chemist, a physicist or your teacher, you will probably find that most of them have never studied Avogadro's original paper and only know the atomic interpretation. Some will find it quite difficult to understand Avogadro's logic and the possibility of holistic division; others will defend the atomic hypothesis as extremely useful, successful and indispensable for science. Still others seem to believe that people of the present are by nature more intelligent than they were in the past. Accordingly, they find it difficult to digest the idea that a man who lived two hundred years ago, when people still traveled in carriages and rifles were still loaded from the front, should have had more profound ideas than many modern scientists. Apparently, they believe in linear progress of science. If this is the case, they are not really aware that knowledge is not inherited but passed on – and in this way also drags along many old, unexamined and often unconscious assumptions. The fact that Avogadro's discovery of the bipartition of gaseous matter remains correct even if CANNIZZARO's atomic interpretation should prove to be wrong does not immediately make sense to everyone. For this reason, it seems almost inconceivable to many contemporaries that the atomistic world view of physics and chemistry could also be just an unfinished painting, like all our scientific knowledge. Actually, all these paintings rather resemble lenticular pictures whose motifs can change abruptly at any time – as soon as the observer moves and changes his perspective. The new face of nature that then appears so suddenly deviates so much from the traditional, seemingly secured world view that the cognitive barrier may seem insurmountable at first glance. This has not least to do with the fact that we – scientists not excluded – are constantly in danger of confusing our models and design drafts, i.e. the constructed *actuality* in our minds, with *reality*. And this has also to do with the fact that body and object conceptions are so essential for our thinking. Exactly this problem is also the core of the interpretation problem of quantum physics, which calls out to us loud and clear: Check your assumptions first, before you despair on a non-understood reality which is obviously not so constituted as you imagine! But, the reflexive component of the natural sciences, which could easily compensate for this, falls far short of the actual requirements nowadays. That is why the hypothetical character of all science and all scientific models is so quickly forgotten once a theory is accepted.

Preview to relax: Can spectral analyses prove molecular cell division?

The discussion of AVOGADRO's hypothesis shows how difficult it can be, even for smart people, to overcome established thought patterns and the imprint by learned knowledge. But if one has studied the original essay, it is no longer surprising that the molecule hypothesis remains valid even if the atom hypothesis is wrong. AVOGADRO deliberately avoided the term atom; it is found neither in the original French nor in the English translation. The word atom occurs only in the German translation of 1902, and there only once, which was obviously a negligent mistake of the translator – and that was certainly not WILHELM OSTWALD, for he doubted the atomic assumption. Today, more than a hundred years later, we again have good

Figure 11: Molecular cell division
Imagine a spherical, homogeneous integral molecule – in this case hydrogen – which divides holistically like a biological cell. Due to the cellular structure, the divided molecule physically remains a whole.

reasons to doubt the atom hypothesis. On reason is the continuing inability of physics to make physical sense of its quantum physical findings and to design a new model of Nature. The other reason is the assumption of a preconfiguration of the molecule. The question is how to understand molecular structure formation when the term atom loses its physical meaning. And to understand this structure formation means to understand the molecular division. Thus, if the integral molecule or an arbitrarily chosen gas volume is not molecularly preconfigured, i.e., is homogeneous, or already has an internal structure but is still holistically divisible, we need a completely new explanation for how a whole can divide, form a differentiated physical structure, and still remain a coherent whole.

From the chemical experiments it clearly follows that a process of division of matter must take place, but this does not force us to conclude that the division must be mechanical in nature, i.e. produce separated parts. Chemical reactions are quite natural processes that also occur independently of human beings and their ideas about the constitution of matter. So if we do not confuse our models with reality, then mechanics, particles and the atomic hypothesis are by no means prerequisites for the validity of chemical reactions, for Avogadro's molecular hypothesis, or for a structure theory of matter. If we give up the ad hoc hypothesis of a bipartite preconfiguration for which there is no experimental evidence, we can reasonably assume that the division must be holistic in nature. Thus, it does not matter whether we are dealing with a homogeneous substance or with an already differentiated matter that can continue to divide holistically – both are possible. To understand this, we only need to think of the biological cell division of an egg cell. And in this simple way we have discovered something really new. In this new picture, the division of an integral molecule generates two coherent daughter cells. This is a new physical model that changes our understanding of nature and reality in a revolutionary way.

Avogadro's elementary molecules (our modern atoms) are then no longer the primary components of molecule formation but the secondary product of molecular cell division. This gives a clear conception of reality: either we accept a divisible 'atom', which however only makes clear that the atom hypothesis is physically senseless, or we accept the idea of an integral molecule that can divide holistically, which makes physical sense and explains the structure formation of nature – for the first time, by the way.

You may now ask: *"Can this really be true? If molecular cell division really exists, there should be experimental evidence for it by now, right? And what about spectral analyses? Can spectral analyses prove molecular cell division?"*

A perfectly legitimate question. A whole series of quantum experiments will show that this hypothesis is indeed experimentally testable, which will be discussed in more detail in the second volume. They will show that cell division and field branching processes are in full agreement with experiment. But in order to dispel your quite justified doubts and not to overstretch your patience (after this really exhausting Avogadro chapter), I will already at this point let you in on a secret that will probably shock you: That we do not need the atomic hypothesis at all to explain the nature of molecules has been known for almost a hundred years – at least to a small circle of experts! In the old atomic-mechanistic picture the molecule consists of two somehow connected atoms, in our new molecule picture of at least two daughter cells. Both are supposed to form a whole. This raises the extremely interesting question of the true nature of *chemical bonding* and whether there is further experimental evidence to support the claim that the atom concept is expendable. The same question was asked as early as 1926 by two young quantum physicists trying to understand the mystery of chemical bonding. They came to exactly this conclusion – inspired by spectral analyses. We will come back to that in the second volume, so I just want to quote their key statement here:

"In general, no attempt is made to treat the molecule as consisting of atoms or ions."

Since then, the circumstantial evidence has been lying in the criminal archive of physics – right next to the yellowed manuscripts of Avogadro. Apparently, leading physicists could not read the tracks, although they led directly to the prime suspect – namely, the atom. This can hardly be reasonably explained because, after all, our two young physicists had taken the experimental clues very seriously and claimed that the notion of atoms was dispensable for understanding molecular structure. They were able to show that the nature of the chemical bond and the structure of the hydrogen molecule – in accordance with the spectral pattern – can be much better explained by the conception of a *dividing atom*! This results in a molecule that is in-itself oppositely structured, yet remains a whole. If that cannot be interpreted as molecular cell division and dimagnetic polarization, I'll eat my hat...

The evidence for this deduction comes from molecular spectroscopy and the quantum orbital theory of the electron. Molecular spectroscopy is an experimental method for analyzing the spectrum of light emitted or absorbed by gases, which provides at least indirect information about their molecular structure. Or as a Crime Comissioner would say: atomic or molecular spectra are the fingerprints of matter that provide hard evidence. Between 1926 and 1932, the results of spectral analyses led to the *Molecule Orbital Theory* (MOT), in which our two young physicists found it necessary to abandon the atomic molecule of Cannizzaro. Based on the electron orbital theory and wave equations, they designed structurally holistic molecule models that were in better agreement with the experimental facts. In this way, they challenged the fundamental assumption of physics and chemistry that molecules consist of individual atoms. In other words, physics has been asking us since 1932 to abandon the atomic concept when explaining the "physique of molecules"!

Obviously, Cannizzaro's atomic molecule model was already ripe for a critical debate around 1932. But such a natural philosophical discussion apparently never took place – and for two reasons (which we will only hint at here, since they will be discussed in detail in the second volume). On the one hand, there was a competing theory of electron orbitals that did not challenge the atomic idea. This was the atomic orbital theory (AOT), which was initially applied very successfully to molecules. Its leading exponent: LINUS PAULING. On the other hand, NIELS BOHR, the great mastermind of the quantum mechanical interpretation, had decided already in 1927, contrary to EINSTEIN's explicit warnings, to categorically exclude natural philosophical questions about the physical meaning of quantum physics from the physics... Oops!

So the failure of the atomic paradigm is not as dramatic as it seems, and certainly not mystical. The strangeness of quantum mechanics has nothing to do with a spooky nature of reality or natural limits of the thinking mind, but with the failure of our greatest physical prejudice. Molecular division shows that the assumption of indivisibility is by no means a physical necessity to explain the nature of matter. It also shows that the molecular nature of gases need not be based on mechanical ideas, i.e., three-dimensional objects moving in a vacuum or through space. And it makes clear that AVOGADRO's hypothesis is not at all a proof of the correctness of the indivisibility hypothesis, but exactly the opposite: The principle of (holistic) divisibility must hold universally for any conceivable theory of matter. The nature of this division process is obviously at the heart of physics, regardless of whether one wishes to opt for substance-based or field-based model conceptions.

Summary: The first switch of cognition

With the discovery of molecular cell division we have found a simple way to replace the failed atomic hypothesis already in classical chemistry and physics with a completely new physical idea that can be tested experimentally. And with this also becomes clear that the failure of the body and atom conceptions in our minds can already be explained by ordinary chemical reactions. And this proves loud and clear that the failure of a physical hypothesis has not the slightest effect on the most important premise of science – that a reality, of whatever kind, must exist. This realization should actually elicit a deep sigh of relief from every quantum mechanic.

Fresnel: The rise of wave theory

Around 1805 Thomas Young discovered the divisibility condition of light that refuted the atomic particle hypothesis for light. Shortly thereafter, in 1811, Amedeo Avogadro discovered a similar divisibility condition in chemistry, which challenged the atom hypothesis for matter as well. However, this was not felt to be conclusive evidence because simply still too little was known about the properties of matter and light. One could avoid this conclusion, for example, by assuming that a light beam consists of many individual light rays which simply separate from each other in the branching process, i.e. simply go apart. In such a mechanical composite-hypothesis, as in Cannizzaro's interpretation of molecule division, there is nothing to be seen of a physical division process. Most researchers were so impressed with the possibilities of modeling reality offered by Newton's mathematical principles that they failed to recognize that the underlying atomistic- mechanistic world view rendered holistic division and branching processes invisible, i.e. natural phenomena that are clearly observable otherwise (look at branching trees, it's physics). In the mechanical world view, branching has no physical meaning and no explanatory potential. It simply doesn't exist in a model universe which revolves about the motion of and the forces between discrete bodies and particles. Thus many physicists could not really accept Young's wave model despite the interference condition. They still hoped to understand interference effects either with indivisible particles of light or indivisible light rays and waves. This problem still exists today.

However, unlike the particle model, the wave model was capable of adequately representing the branching behavior of light in double-slit experiments and other optical phenomena. This did not yet mean that Young's wave model was already perfect and the analogy to water and sound waves was necessarily correct. But at least waves in fluids and sound waves in gaseous substances have no problem to pass two openings at the same time and create two secondary waves and superposition phenomena. This is exactly what a tiny particle cannot do: If it were to split mechanically, only a double light spot would appear on the detecting screen, but no interference pattern. There was simply no way to get around the divisibility condition, so the only option was the wave model.

To gain more clarity about the nature of light, the French Academy of Sciences announced a Grand Prix in 1817 for the best explanation of the manifold appearances associated with the behavior of light. The aim was to understand how light rays, the branching of light rays, the colors of light, the speed of light, and the phenomena of dispersion (distribution of light) including refraction, diffraction, and polarization come about. Given the diffraction (bending) and superposition effects of light on double slits, fine gratings, and obstacles that could only be explained by wave theory, it was hardly surprising that there were no submissions from particle physicists.

AUGUSTINE FRESNEL (1788-1827), a French civil engineer and science amateur, deeply interested in the nature of light and optics for pure pleasure, submitted a self-developed mathematical wave theory, apparently without knowing YOUNG's double-slit experiment and wave theory. Gifted with exceptional mathematical skills, FRESNEL developed sophisticated mathematical equations that described the behavior of light in diffraction, refraction and interference processes in great detail. As in YOUNG's model, light should consist of *longitudinal* waves in an elastic medium like sound waves. After passing through a small hole or slit, the wave expands in all directions in the form of a hemisphere. At the edges of the obstacle, the primary light wave locally induces many new tiny secondary waves, which in turn expand in the form of spherical or hemispherical waves. But all these new hemispheres – or spheres – remain within the envelope of the original spherical surface of the primary wave. This was the application of HUYGEN's principle, which now had to be combined with YOUNG's superposition and interference effects.

The contact with matter thus causes a spherical light wave to form myriads of tiny spherical partial waves. If one illustrates this picture with light rays (or wave normals), one can recognize branching processes in it. In other words: in contact with matter, the light somehow structures itself more finely. HUYGEN's principle says that this structure formation takes place within a holistic system that always remains enveloped by the original spherical wave. This is reminiscent of a fractal process (a term that did not exist at the time) and already gives the impression that the diffraction of light may well

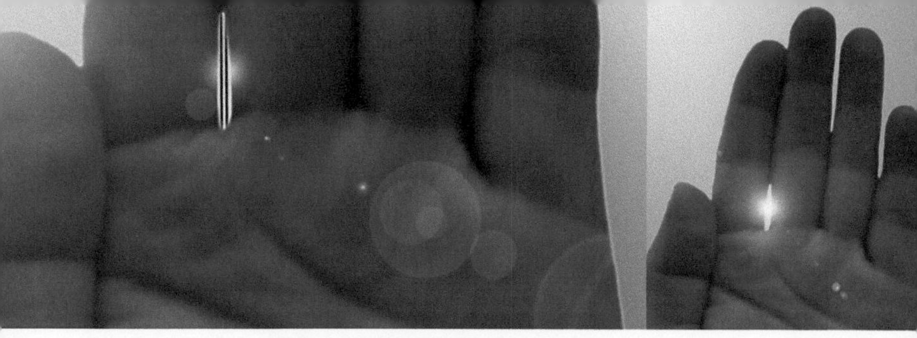

Figure 12 Fresnel's single-slit pocket experiment

Fresnel (pronounced 'Frenel') left us not only a mathematical wave theory of light, but also the simplest experiment in the world. It goes like this: Take your hand out of your trouser pocket and hold it against the sunlight in such a way that some light falls between two fingers through a small gap. If you look closely, you can see with the naked eye a dark line in this gap, which stands vertically like the string of a harp. On closer inspection, this line appears to divide further into two, four, eight lines, and so on. These are interference stripes, where the light is completely cancelled out because of reflection and diffraction at the two fingers. Apparently, then, we can even see things that emit no light at all! This experiment is so simple that no technology is needed. Even a stark naked man (or woman) on a desert island could perform it.

have something to do with cell division and branching processes, as suggested by the double-slit experiments of quantum physics and partial reflection: If one imagines a light ball which does not move at all, the ball differentiates inward like an egg cell which divides holistically finer and finer. This is an illustration of HUYGEN's principle, but also an interpretation – because at that time nobody could know anything about cell division processes. As the word *diffraction* already reveals, at that time people thought in mechanical analogies of a breaking up of light into different partial rays or wave trains, i.e. a splitting of light into its components. Since the light rays change their direction during these processes, this term also means a bending and deflection of the light rays or wave normals.

FRESNEL realized that light rays or waves, due to the splitting at the edges of an obstacle, can bend around the obstacle, shifting slightly against each other and thus superimposing themselves, resulting in the same interference effects that were already known from water and sound waves. Depending on the displacement of the wave crests and troughs, the waves cancel each other out or reinforce each other. In this way it became again clear that the particle picture could not apply to light. In order to demonstrate diffraction and interference effects anytime and anywhere, FRESNEL invented a simple "experimental setup", the famous single-slit pocket experiment (Fig. 12).

At the academy, however, there were not a few scientists who could hardly imagine that Newton's theory could fail because of the true constitution of reality. Among them were the famous astronomer Laplace and the great mathematician Poisson, who immediately vetoed Fresnel's wave model. Poisson himself calculated a solution from Fresnel's wave equations which was to immediately disprove the wave theory: If the wave theory were correct, a light beam should be able to bend around an obstacle, e.g. around a small disk. In doing so, the incident light beam (or primary wave) would have to split into many individual light rays or secondary waves that bend slightly inward as they flow around the obstacle. And this in such a way that they would all have to reunite at a certain point far behind the disk. If this would be correct, at this place, in the middle of the core shadow, a tiny luminous point should appear. Poisson, a staunch follower of Newton, believed that a theory that predicts something so nonsensical simply must be wrong (Fig. 13).

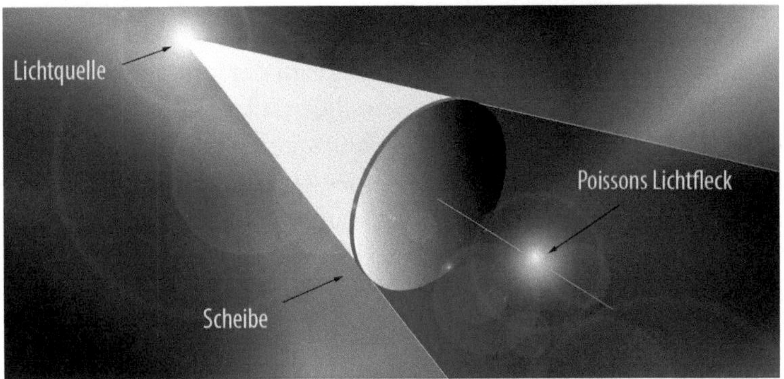

Figure 13 Poisson's light spot

But Newton had left another important message: Every hypothesis must prove itself in the experiment! So Arago, the chairman of the jury, devised an experiment, which confirmed the existence of the predicted light spot and thus Fresnel's wave equations. And so, for the first time, wave theory gained widespread scientific recognition. Newton's atomistic idea was experimentally refuted for the second time, at least for light. The particle model was thus almost finally off the table, and the wave theory of Young and Fresnel was taken very seriously from now on.

What we can see in this episode is that Poisson's light spot perfectly describes what happens to individual light quanta, light rays or light waves when they bypass an obstacle. At the double slit, it is the narrow vertical bar between the two slits that forces the light to divide, just like at a tree (p. 29 Fig. 5 right and Fig. 6): A quantum of light behaves practically like a light beam consisting of only one light ray or wave, which can branch out and reunite at one point! If this is independent of the light intensity, the Poisson formula, so to speak, already describes the elementary process of branching and reunification of light rays which quantum theory cannot accept as reality. The attentive reader naturally asks at this point:

"How can it be that one can model this situation with Fresnel's wave theory without problems, but not with the wave equations of quantum theory?

The most important reason is probably that the wave theory, which Fresnel had originally developed, was changed, and that by Fresnel himself. The wave equations which Poisson had used to calculate the light spot were based on the idea of longitudinal waves. The wave theory Fresnel established few years later were based on the idea of transverse waves. This transverse wave concept was adopted by all subsequent wave theories: Maxwell took it over from Fresnel, Lorentz and Einstein took it over from Maxwell, and De Broglie and Schrödinger used it for electron waves. So what is the difference between transversal and longitudinal waves?

Transverse waves oscillate transverse to the direction of propagation, i.e. always perpendicular to the direction of the light ray (e.g. to the right and left, or up and down like water waves). In contrast, *longitudinal waves* oscillate in the direction of propagation, as is known from pressure and sound waves (i.e. forward and backward). Under resonance conditions, they can also form standing waves that cannot propagate, as in the resonating bodies of musical instruments. These are *stationary vibrating* longitudinal waves in a gaseous medium. In the case of sound waves, this medium is air; in the case of light waves, the medium of the oscillations was assumed to be a still unknown gas, which was called *aether*. These stationary waves can produce coherent, complexly structured three- dimensional shapes and forms. The complexity or depth of structure of these vibrational forms depends on the frequency, i.e., the vibration energy supplied, the nature of the medium, and the geometry of the resonance cavity. The simplest forms of oscillation of standing waves shows a tightly stretched string. Depending on the energy input, the vibrating string forms one, two, three or more wave sections or vibrating bellies, which are separated from each other by nodes. The nodes do not vibrate because the waves cancel each other out at these points.

Although people were not thinking in this direction around 1826, the theory of self-structuring standing waves, which originally came from the mathematical-physical analysis of the resonance properties of musical instruments, proved to be very valuable a hundred years later: It was the best way to understand and model the quantum-like structure of atoms and molecules. Only the question, what should actually vibrate there and form the mass particles (molecules), could not be clarified until today. Is it stationary, oscillating light, from which the molecules and the matter emerge? Or is there still a so far undiscovered gas in the universe which is the medium of the light waves? Could this gas produce stationary longitudinal waves under still unknown resonance conditions, which are reflected in themselves and therefore can interfere with each other? Could stationary matter waves be formed in this way, which are complex structured, extremely condensed, and inherently stable? Or are we dealing here with invariant structural principles of Nature, which always create the same chemical elements under the same conditions?

So, before FRESNEL moved to the transverse wave model, light was assumed to be made up of longitudinal waves, just like sound and pressure waves in the air and underwater (not to be confused with the waves at the surface of the water, which are vertically oscillating transverse waves). The purpose of introducing transverse waves was solely to explain the polarization of light rays. Until then, YOUNG and FRESNEL had no problem performing interference experiments with direct sunlight, which is unpolarized, and to explain the nature of light with longitudinal waves. However, the polarization of light forced them to assume a new physical degree of freedom for light rays that had interacted with matter. In mechanics, the term *degree of freedom* refers to the possible directions of motion of an object. For example, the piston of a cylinder can only move back and forth in one direction and therefore has only one degree of freedom (of motion). The same is true for longitudinal waves: Since they represent pressure waves in gases and liquids, they can only oscillate back and forth in the direction of propagation and therefore also have only one degree of freedom of motion. In general, this term can also denote new physical properties. To a mechanic who handles particle-like quanta in his mind, this must seem obscure by nature, since he cannot imagine anything else besides his mechanical analogies.

The first experiments conducted jointly by FRESNEL and FRANCOIS ARAGO were concerned with the behavior of branching sunbeams in glass bodies and polarized partial rays that appeared to emerge from such processes. Many of these findings were originally based on the longitudinal wave concept. It was only 1821 that FRESNEL, inspired by THOMAS YOUNG, began to replace this conception with transverse waves. It is probably no exaggeration to say that these experiments and the mathematical equations derived from them laid the foundations of optical wave theory within just a few years. They always involved – you may have guessed it – the branching of light beams producing polarized light rays. FRESNEL specified the rules of interference, described birefringence in crystals and glass (double refraction), which not only produced two polarized partial rays but also seemed to slow down one of the two rays, and formulated the rules of partial reflection, which also resulted in two polarized rays with precisely calculable intensities. His mathematical wave theory almost completely explained diffraction, refraction, refractive index and the relationships between wavelength, frequency and material properties. This is, of course, only a short summary of FRESNEL's great work, which was produced in only eleven years (he died at the age of 39). Many readers may also be familiar with the FRESNEL lenses that he developed for lighthouses as the first technical application of his findings on the refraction and reflection of light waves.

However, despite thorough analyses, FRESNEL failed to understand the true nature of polarization, as we know today. This is the key problem that quantum theory still struggles with two hundred years later, preventing us from grasping the true nature of light and matter. Thus, if we want to understand quantum theory and its cognitive problems, we must first of all understand the nature of polarization. For this reason, we should take another look at the history of science and critically re-examine the reasons for the conceptual shift from longitudinal to transverse waves.

Polarization and the introduction of transverse waves
The polarization of light rays was discovered and researched around 1809 by MALUS (1775-1812). He had found that reflected sunlight in a double-refracting prism of quartz splits into two rays of light as in a thick Iceland spar, which behave differently from normal sun rays: One goes straight through, the other is double refracted. This phenomenon, called *birefringence* was already known to HUYGENS. But nobody knew why *two* light rays appeared – there was simply no dynamic model for it. The only thing that could be done was to take note of the existence of two light rays and to describe their movements kinematic without going into causes and forces. Apparently, a beam of light could split or branch into two partial rays, one of which behaved 'ordinary' and the other 'extraordinarily'. However, when MALUS rotated the prism step by step by 90°, he found that the two rays alternately disappear. The ordinary light ray was apparently as extraordinary as the other, even if it was not refracted. The branched rays had apparently acquired a new physical property, which leads to different diffraction effects in optically active crystals. MALUS called this new property *polarization* and concluded that the resulting partial rays must be polarized oppositely. He also discovered that all sunlight reflected from smooth surfaces such as lakes, mirrors, glass panes, metal sheets and wooden surfaces is polarized. He concluded that this new physical property was due to division and reflection processes and proposed 'polar' properties of *light particles* as an explanation, which he illustrated with magnetic monopoles (MALUS was a follower of NEWTON's atomic particle theory). This was a brilliant analogy and discovery of fundamental natural philosophical importance. At that time, however, nobody could know that this proposed analogy *actually* hides dipolar, oppo-

site magnetic properties that a light beam acquires by branching. This was not discovered until 1845 by FARADAY.

FRESNEL used another crystal to determine the polarization of light rays: He sent a light ray, already polarized by reflection, through a thin, translucent lamella of tourmaline (another crystal with aniso-tropic or chiral structure) and found that the intensity of the trans-mitted light depended on the angle of rotation of the lamina. If one turns the tourmaline disc around the axis of the incident light ray, the transmitted light is dimmed further and further until no light passes through at all. If one turns the disc even further, the intensity of the transmitted light ray increases again until it returns to its full extent in the initial position. The dependence of the transmitted light intensity on the angle of the analyzer is described by MALUS' law. The zero-degree position is defined as the position at which all light is transmitted (we return to this topic in the second volume by discussing in detail the polarization experiments with single quanta of light and illustrating them vividly with figures).

Apparently, branched and polarized sunlight has anisotropic prop-erties, as do some translucent crystals and ordinary double-refract-ing glass panes in partial reflection experiments. This means that branched light cannot be homogeneous anymore, but must have a chiral structure. The angle of the tourmaline crystal apparently did not matter with unpolarized light, because that passed through in any rotation angle. The relative orientation between the tourmaline lamina, i.e. its directed crystal structure, called *optical axis*, and the incident light ray mattered only with polarized light. Consequently, a polarized light ray had to have a 'handedness' (a left- handed or right- handed character). This is not an expression of a spatial direc-tion, but of the ontological mirror symmetry of a bifurcated light beam. A partial ray seems to have a preferred direction only because the extent of its absorption is related to the angle of the optical axis. This is at least my interpretation. The question that YOUNG, FRES-NEL, ARAGO and other researchers asked themselves was: can polar-ization be explained by longitudinal waves? How do normal, direct sunbeams acquire dipolar properties? How can the wave theory be adapted to these facts?

The simplest possibility would have been to conclude that individual longitudinal waves, i.e. pressure waves in a gas, can branch and thereby produce dipolar properties of the partial waves. In acoustic theory, however, this would mean that the gaseous medium itself branches and takes on mirror-symmetric (anisotropic, dipolar, chiral) properties. In fluid mechanics, which at that time existed only in the rudimentary form of BERNOULLI's (1700–1782) kinetic theory of gases and hydrodynamics, this would mean that individual streamlines could branch mirror-symmetrically and take on dipolar properties. In the last consequence that would mean that even pressure and sound waves can cause molecular divisions – what seems to contradict our present conception of the molecular structure of matter. But this idea already touches the topic of a new kinetic theory of matte, to which we will turn later.

In other words, branching processes would transform the homogeneous gas into an asymmetric, anisotropic medium, which in turn would produce feedback effects on the propagation of disturbances (both acoustic pressure waves and longitudinal light waves). For light, this would practically mean that branching processes of the hypothetical aether gas would transform the homogeneous, translucent gas into a gas of chiral nature with an optical axis. Of course, this can be applied to the entire molecular structure and to individual molecules. The medium and each single molecule would then acquire an axis of symmetry as known from optically active crystals, which would also have to change the refractive index. In this picture, then, we recognize the molecular cell division processes that we described in the Avogadro section. All these considerations show how closely the theory of light was originally connected with the theory of gaseous matter, and that the question of holistic division processes of gas volumes does not only arise in AVOGADRO's molecular theory. But this line of thought was not pursued at the time.

FRESNEL worked his way up to a different conception step by step: What was particularly remarkable was that polarized light behaved differently from normal or direct (i.e. not reflected) sunlight. Sometimes polarized light beams could be made to interfere, sometimes not. This depended on the relative polarization of the two light rays, but also on whether they were partial rays that had emerged from the branching of a light beam. With the help of a tourmaline disc it was apparently possible to analyze in which spatial direction the polarization plane pointed. Since this picture was so sensual, FRESNEL began to associate the dipolar properties of light with a plane of polarization. The dipolarity of two partial rays resulting from the branching of an incident beam of sunlight was expressed by the fact that their planes of polarization were always oriented at right angles relative to each other. So FRESNEL began to speak of perpendicularly or orthogonally 'plane' or 'linear' polarized light waves. As he found out, they cannot interact in this state – unless the original ray was already polarized. So in this case, only *one* branch of light was further branched. In addition, there was another form of polarization in which the polarization planes of the two light rays pointed in the same direction, what he called parallel polarization. In this state, the two polarized rays can interfere. In 1817, a more thorough investigation by FRESNEL and ARAGO showed that four polarization cases can be distinguished:

1) Two plane or linear polarized light rays or waves can interfere if their planes of polarization are parallel (or anti-parallel).

2) Two light rays of different origin polarized at right angles to each other represent opposite 'light forms' which cannot interfere. This leads to the well- known statement that light rays or waves, which originated independently of each other, can pass through each other without interfering.

3) Two light rays of the same origin polarized at right angles to each other, as produced by the bifurcation of an unpolarized light beam (sunlight) during double refraction, cannot interfere.
(From this we can conclude that the branching of sunlight on the double slit or on a tree must produce parallel or antiparallel polarized partial rays, because they can interfere).

4) The only exception to case 3: If the two orthogonally polarized partial rays have emerged from the branching of an *already polarized partial ray*, they can be recombined and *interfere*.

In the longitudinal wave theory of Young and Fresnel, however, it was unclear how to reconcile the dipolar properties of light, which geometrically appeared so vividly as polarization planes, with longitudinal oscillations of shock, pressure, or pulse waves. If polarized light rays were still longitudinal waves, the rotation of the tourmaline lamina should have no consequences. So to reconcile the wave model with the phenomenon of polarization, it was necessary to add another *physical dimension* to one-dimensional light rays or longitudinal oscillations. But how should one imagine this?

Theoretically, this could also be explained in the following way: If a light ray were not a one-dimensional line but a band, a two-dimensional plane, then this plane would have a certain direction in space. The intensity of the transmitted light would then depend on the relative angle between this plane and the optical axis of the crystal. If the optical axis and the plane are parallel, all the light is transmitted; if they are perpendicular, the light is blocked. Conclusion: the wave must oscillate in the plane of the band, that is, transverse to the direction of propagation of the light ray. Thus, polarized light rays can no longer be longitudinal waves, but only transverse waves. Of course, this assumption of a band-like nature of the light waves was pure fantasy at first. It was merely a constructive theory, a model that could correctly describe the dependence of the intensity of the transmitted light on the direction of the optical axis of the crystal. Because the mathematical model was so successful, the imaginary wave plane took on more and more the character of a physical reality. At this point one can see how the confusion of a constructive physical model, which may well reflect the existing relationship, with reality begins. However, the fact that this model cannot apply to reality turned out only in the context of quantum physics.

So it was the inability to explain the dipolar properties of branched light rays with longitudinal waves that forced Fresnel and Arago to gradually replace the idea of longitudinal waves with transverse waves between 1817 and 1821. They corresponded by letter with Young, who had apparently been the first to suggest that polarization could best be explained by transverse waves. But Fresnel had recognized three other significant facts:

First, that the two partial rays produce complementary colors in the case of double-refraction (birefringence). This means that ontologically they must indeed be complementary in nature. Second, that the two polarized partial waves are geometric mirror images of each other, which is mathematically expressed with a 180° inversion of the wave phase (where one wave has a belly, the other has a trough). And third, that the complete reflection at a glass pane causes an inversion of the reflected ray, which must also be described with a 180° phase inversion. However, this cannot only describe a simple rebound as in an elastic impact or an optical reflection of the ray, which would then actually have to remain unchanged. To make it even clearer: A tennis ball bouncing off a wall remains the same tennis ball; it only reverses its direction of motion. This should actually also apply to light rays and waves. But here we have to do with a "tennis ball" or ray of light which completely inverts itself during reflection, that is, turns its inside out. Of course, this immediately reminds us of the 'inversion transformation' of a glove: if you turn it inside out, a right glove becomes a left glove and vice versa. This is exactly what happens in the reflection of light: the reflected ray is the ontological mirror form of the incident ray. And only if one considers both together, one recognizes that they represent a physical system which is of mirror-symmetrical nature. In other words, all three processes can actually be understood as mirror-symmetric branching processes, producing left- and right-handed partial rays or enantiomorphic field structures. And if the unidirectional definition of the speed of light would not exist, one could directly assume a standing wave or field branching process, which starts from a node and branches in both directions at the same time. If this is true, real structural changes of the light or the 'electromagnetic field' are hidden behind the wave theory.

Polarization and the composite-hypothesis of light

FRESNEL, however, pursued the following idea: The simplest explanation of polarization would be to take the polarization plane physically seriously. This would mean that light waves always oscillate in a plane *transverse* to the direction of propagation. This plane of polarization already exists when the waves are emitted. According to this conception, a sunbeam is always a bundle composed of many individual transverse waves, each oscillating in a particular plane. The individual light rays in a light beam are therefore always polarized, whereby the direction of polarization is random. The different spatial orientations of the polarization planes would average out so that the entire beam of sunlight appears unpolarized. A polarization analyzer or polarization filter like a tourmaline or nicol lamina would then filter out all the individual waves that do not have the correct orientation of the wave plane; they would therefore not be transmitted. The transmitted light beam then consists of transverse waves that all have the same plane of polarization. In this way, FRESNEL's transverse wave model explains how polarized light waves or rays come about: They are practically already there from the beginning and are only filtered out and sorted.

At this point, however, it also became clear that the theory of light should always yield the same result – in practice and in theory – regardless of the brightness (aka intensity) of the incident light ray. In the double-slit experiment, for example, the typical stripe pattern would always have to appear on the detecting screen. If the light were very dim, the pattern might not be seen, or it would be necessary to collect the light energy with a light-sensitive film and expose it for an appropriately long time. However, the invention of light-sensitive photographic plates was to take several years – the first were invented in 1826 and 1837. In fact, the first double-slit experiment with a photosensitive detecting screen did not take place until 1907, when TAYLOR wanted to test the validity of the divisibility condition even at the lowest light intensities and thus EINSTEIN's quantum hypothesis of light. The result: The interference stripe pattern formed even at the lowest intensities; consequently, the divisibility and interference condition – the simultaneous passage of the double slit – had to be valid even for the smallest amounts of energy down to single emitted light rays or energy quanta.

So if the intensity of the beam of light is very low, the question is only how long we have to wait until enough impact points are registered for the interference pattern to become recognizable. Then, in the case of individually emitted light rays, the interference pattern becomes a random pattern of individually registered points, just as if we were observing a normal double-slit experiment with continuous rays in extreme slow motion (low light intensity = long exposure time). Thereby we notice that the apparently continuous radiation acts in the form of single light pulses or light quanta, although the radiation must have branched *before* at the double slit. So it looks as if the wave theory could represent branching processes without problems. However, if one looks more closely, it becomes clear that it is impossible to explain dipolarizing branching processes with the *usual* transverse wave theory, since the explanation of polarization – and thus the nature of transverse waves – proposed by Fresnel excludes branching processes from the outset:

"The act of polarization consists not in creating transverse motions, but in decomposing them in two fixed, mutually perpendicular directions, and in separating the two components." [12]

Attention! Exactly this assumption of an additively composed mechanical system leads hundred years later to the problem of understanding the true nature of polarization, which reappears in quantum physics as "spin" of virtual particles. It took a long time until it became halfway clear that this "spin" can neither be an objectively preexisting physical property nor a rotational motion of particles, but must be an *emergent* physical property, which only arises in the course of an interaction – which, according to our new hypothesis, is a polarizing holistic division and branching process.

The cause of this cognition problem is actually easy to see: In Fresnel's wave model, a single light beam must consist of at least two transverse waves, which, moreover, must be oppositely polarized in order to be divisible in this theory at all! So it is Fresnel's composite hypothesis which axiomatically excludes a reasonable theoretical treatment of single light rays, waves or 'light quanta' as soon as they branch at the double slit, a glass pane or a polarizing filter. In other words, the transverse wave theory rules out branching processes of single light rays and waves from the outset. But this contradicts the law of optics, according to which the behavior of light cannot depend

12   Fresnel 1821 (quoted by Buchwald): https://en.wikipedia.org/wiki/Augustin-Jean_Fresnel

on its intensity. All this makes clear that the transverse wave model was only a constructive hypothesis invented to explain the dipolar properties of branched light rays as given, although they become detectable only after division. This immediately reminds us of the ad hoc hypotheses of AVOGADRO and CANNIZZARO, who assumed a preconfiguration of the integral molecules *due to their divisibility*. For the same reason, FRESNEL assumed a preconfiguration of light beams *due to their divisibility*, which should explain polarization at the same time. Incidentally, the fact that molecules actually exhibit the same mirror-symmetric branching and polarization behavior was not known at the time. So 1821 was the decisive moment when the longitudinal pulse and pressure wave model of HUYGENS and YOUNG transformed to the transverse wave model of FRESNEL, which was later assimilated by MAXWELL's electromagnetic theory.

In this way, the other possibility that the polarization of light could be due to physical branching processes – like every structure formation in Nature – was already excluded from the thought spectrum of the wave theorists two hundred years ago. And as it looks at the moment, nobody has thought about this possibility yet. However, if the complexity of the behavior of light in interactions with matter is stripped down to the essence of an elementary process, one can easily recognize a dipolarizing branching process. And HUYGEN's principle, supplemented by the coherence and interference principle, already makes it clear that a holistic view of light is well within the bounds of what is physically conceivable.

Summary: The second switch of cognition

This short history of wave theory shows that our suspicion that wave theory knows no real branching processes at all is indeed well founded. The model of transverse waves cannot represent branching processes because it follows the mechanical idea of a composite system. Therefore, the theory cannot map the branching of single light rays and light energy quanta. This is, so to speak, the blind spot of wave theory, the infamous 'theory glasses' whose slits of vision block out everything else. And in this way we have discovered the next switch of cognition in the history of physics: Fresnel's findings may well be interpreted as genuine physical branching processes that produce dipolar, i.e. ontologically mirror-symmetric, properties.

Oersted: The discovery of electro-magnetism

By 1820, there was already enough evidence to argue against indivisible light particles in favor of wave theory. And yet it was to take many more decades before the majority of natural scientists could accept that still another world existed besides the Newtonian world view, which could not consist of atomic particles. The delay came about because it was not yet known how to understand electricity and magnetism and what they have to do with the dipolar physical properties of light. But with the recently discovered wave properties of light and newly emerging electrical and magnetic effects, more and more phenomena appeared that could no longer be understood with the body concept of mechanics and directly acting forces. But some researchers were still not convinced that the wave concept is really needed. One of them was AMPERE, who had supported AVOGADRO's molecular division concept a few years earlier with similar considerations. Now he set out to design a theory of electric current and magnetism and to explain it on the basis of NEWTON's theory.

Until then, electricity had always been thought of as an atmospheric effect, as if it were an elastic fluid with electric charge that behaves like an invisible gas. The naturalist DU FAY (1698-1739) had already discovered the antisymmetry of electric charges in 1733. His friction experiments with different materials led him to the conclusion that two different, exactly opposite *qualities* of electric fluida exist, which he called vitreous (today: positive) and resinous (today: negative), since they came from glass and amber, a hardened tree resin. Today we are so used to the names *positive* and *negative* electricity that we can hardly imagine that these terms could be completely different as long as they express the same thing: A dipolar property, so exactly opposite electric qualities. Plus and minus, black and white, yin and yang or blue and yellow would also do it – the main thing is that they always mean genuine anti-parts. What these dipolar parts are called does not really matter. One could also call them right and left in a topological sense, as we do for our hands, which are spatially mirrored but nevertheless physically coherent parts. If we turn the inside of a right glove outwards, we call this topology *left*, since only then it fits the left hand. So this relationship and distinction between left and right means no direction in space, but a relative, spatial mirror symmetry, which is a dipolarity or antisymmetry. But since we call the electricities traditionally positive and negative, we may stick to them, as long as we are aware that these are *qualitative* anti-parts and not quantitative terms.

One can easily produce electricity by friction of two different materials, so it was thought that electricity must somehow be contained in it, that matter could be pre-structured and composed by positive and negative charges. Normally, matter appears electrically neutral, so the positive and negative charges must somehow cancel each other out exactly, if they are already there. That does not say anything about the quantity of these fluids or charges within an electrically neutral structure, $+50$ and -50 together result in zero-symmetry as well as $+2$ and -2. By friction of two suitable materials, an electrical charge separation will be created, which generates *static electricity*. This envelops the charged body like an invisible cloud or a gas with an electrical charge that can be positive or negative.

At this point, we can ask whether electric charge separation can also be understood as a holistic division and branching process, which generates electric dipoles, i.e. electric polarization. If so, a potentially 'electric' gaseous fluid would have to structure itself into opposite parts. Indeed, an hard rubber rod rubbed with a cotton cloth subsequently exhibits an electrical charge, so an originally neutral fluid must have divided. One half of the charged fluid remains in form of an electrical charge on the rubber rod (let's call it negative), the anti-part (positive) probably flows off over cotton cloth and human hand into the ground. Of course, these designations could be used the other way around – that is pure convention. If you now touch with the electrically charged rubber stick a piece of metal wire from which a small folded metallic foil is suspended, the two metal strips unfold as if they were branching. The cause of this effect is that two paths are now available to the electric charge. This is immediately reminiscent of YOUNG's double-slit experiment and FRESNEL's theory of light, because it follows that the negative electric charge or fluida must have been evenly distributed to both strips of the foil, i.e., the charge must have divided again or branched further. Figuratively and operationally speaking: We are dealing with a branched

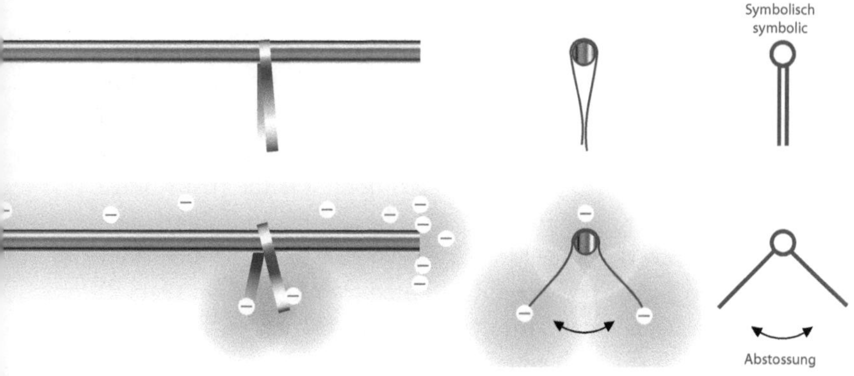

Figure 14 Electroscope

electric system that originally consisted of a negative and a positive branch. These are electric charges that normally would attract each other. But the negative branch now branches further into two negative twigs, which then repel each other. So when a negative electric charge divides or branches again, large repulsive forces are generated (same charges repel each other). The negative electric charge of the rubber rod acts as an initiator, the two spreading metal foils as an indicator. They indicate the strength of the electric repulsion force and therewith indirectly the amount of electric charge. This works the same way for positive charges. This design principle was used for a long time to detect smallest electrical charges with a device called *electroscope* (Fig. 14).

This experiment gives a vivid idea of holistic division and branching processes of an originally neutral substance, through which electric charges and forces are created in the first place. So if you look at the process this way, you don't have to assume that electric charges are already present in matter and somehow neutralize before they are separated. This gives the same picture for dipolar electric properties as with polarized light and molecules: Just because something can divide does not mean that it was already made of parts before.

All this is electrostatics, i.e. the charges or electric fluids stand still; they remain and do not move. No current flows until a short circuit and thus an electrical discharge is generated. In the discharging process, opposite electric charges reunite and make light-sparks or 'electromagnetic' radiation, as should turn out later. The separated electricity could also be stored in an electric condenser or capacitor, called *Leyden jar*. It consists of two oppositely electrically charged metallic surfaces (tin foil) which are separated from each other by a non-conductive glass vessel. This glass body was the first storage device for electricity, invented in 1745. The advantage of this dielectric capacitor was that the electricity accumulated in the molecular structure of the glass body could be stored for months and years (!) and could generate very high voltages during discharges. The disadvantage was that the high voltage could be life-threatening and that the release of the electric current could only be abrupt, unless the current was passed through a water pipe that offered a certain electrical resistance. Since then, it was possible to perform experiments with stored electricity.

But in 1800, ALESSANDRO VOLTA (1745-1827) invented a new storage device called *electric pile*, which was the first electric accumulator or battery. An electric battery generates and stores electricity through electrochemical processes and consists of several elements, called galvanic cells (in the sense of spatial subdivisions). In VOLTA's battery, the galvanic element consists of two metal plates, zinc and copper, which assume opposite electrical polarity in a electrical conductive aqueous solution (salt water or diluted sulfuric acid). The zinc plate becomes positively charged and the copper plate negatively charged, which creates an 'electrical tension' or 'voltage' between the plates. If several galvanic elements are combined in a row, the individual voltages add up. The two contacts where the voltage can be tapped are called the positive and the negative pole of the battery. The electrical charge separation in the battery cell is based on molecular processes. If one connects the two electrically opposite poles of the battery, plus and minus, with a conductive wire such as copper, brass or silver to establish an external circuit, the wire heats up if it is thick enough. If it is too thin, it can become so hot that it melts. At the same time one can notice that the battery discharges, i.e. the electrical voltage decreases. VOLTA concluded that opposite electric charges flow from one pole to the other, forming a constant, uniform and "direct electric current". He was also the first to make a distinction between 'electric tension' and 'electric current'.

If there is no discharge, i.e. no external conductive contact between the two opposite poles or charges, one speaks of electrical tension or voltage. If the two poles are connected to each other in an electrically conductive manner, a short-circuit occurs as with the Leyden jar. However, if the resistance of the wire is high enough or if there is a medium with high electrical resistance between the two ends of the wire, the electricity is continuously discharged and converted into heat. The electric current was imagined as a directed transport of charged particles or volume elements of an electric fluid. However, the electric current was supposed to consist of *two opposite electric currents*: Positive charged electric particles or volume elements of the electric fluid moving from the positive to the negative pole and negative charged electric particles or volume elements moving from the negative pole to the positive pole. In the wire or on the surface, they flow past each other in opposite directions, causing the wire to heat up somehow. The heating was imagined as an "electrical conflict" that had to be caused either by the extinction of opposing electrical charges or by as yet unknown friction motions in matter.

In 1820, the Dane HANS OERSTED (1777-1860) discovered during experiments with such a voltaic battery that a current-carrying wire causes a magnetic action in vicinity which deflects a compass needle in such a way that an angle of 45° is formed between the direction of the wire and the magnetic needle. This led him to the idea that electric charges move in a spiral around current-carrying conductors and that the *motion* of electric charges makes magnetic forces that act on the compass needle. He also found that the deflection of the compass needle weakened regularly with increasing distance from the wire and that the magnetic action was also depended on the intensity of the current. Shortly afterwards, ANDRE-MARIE AMPERE (1775-1836) found out that the angle between the direction of the wire and the magnetic needle is exactly 90° if the influence of the Earth's magnetic field can be compensated with a suitable device. The needle therefore aligns itself exactly transverse to the wire and the electric current. Consequently, a homogeneous, tube-like magnetic state or field had to have formed around the current-carrying wire. However, the idea that fields are independent physical entities which exist in real terms (just like gases and solid matter) was not introduced by FARADAY until twenty years later. Otherwise, fields are mathematical-spatial continuum descriptions of distributions of certain parameters and magnitudes in substances, e.g. of temperatures, pressures or forces. Compared to permanent magnets, however, the electrically generated magnetic state had a rather strange property: Apparently, it had neither a north nor a south pole, but still indicated a preferred magnetic force direction, i.e. it had a certain magnetic polarity. Consequently, the magnetic field tube had to be mono- or homopolar. This can be verified experimentally, as demonstrated by OERSTEDT and AMPERE (Fig. 15):

If you place a compass flat on the current-carrying wire, the needle will always point in the same direction – exactly at right angles to the direction of the wire (if the electric current is strong enough, the magnetic field of the Earth no longer needs to be compensated to prove the transverse direction of the magnetic field or force lines). Thus, the *electrically* induced magnetic force always acts transversely to the longitudinal direction of the wire. If you slide the compass tangentially (or longitudinally) on the surface of the conductor, or pass the compass once around the conductor at a constant distance, the direction of the magnetic needle will not change. That is, the magnetic polarity of the field is always the same. There is no magnetic pole on this ring from which the needle tip could be attracted. Nevertheless, the compass needle always points in the same direction. So how does the magnetic force come about here? Apparently, the electric current – however it is to be explained – creates a polarized magnetic field tube or 'field state' around the wire corresponding to a cylindrical magnetic monopole.[13]

The magnetic field is strongest at the surface of the conductor and decreases radially – like gravity – quadratically with distance. Conversely, this can also be interpreted as radial compression of a magnetic field – or the aether gas – around the conductor. The polarity of the magnetic field depends on the electrical polarity of the wire:

13 Following the ideas Faraday developed later, textbooks usually speak only of ordinary magnetic fields. The insight that electrically induced magnetic fields must always be polarized and consequently can only be the components of a holistically divided magnetic field is therefore not exactly obvious. Only in quantum physics one came across this problem again.

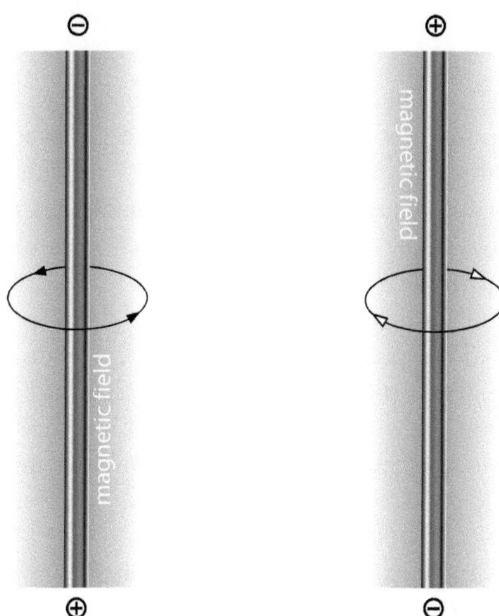

Figure 15 Oersted's discovery of electro-magnetism

A current-carrying metal rod or wire makes a magnetic field hose with a certain polarization around the conductor. The circle represents the path on which a compass is guided around the conductor. Ideally, the bearing axis of the compass needle always points radially to the wire. The circle also marks a cross section through a cylindrical field and a line at which the magnetic force is always equal. Such imaginary lines were later called field lines. The black arrowhead indicates the direction in which the tip of the compass needle (which normally points north) points along the circumference. Seen from the positive pole, this could be called a right-handed magnetic polarization. If you reverse the polarity of the electrical conductor without moving it or the compass, the tip of the compass needle turns in the exact opposite direction. Accordingly, a magnetic field is created around the conductor which has an opposite polarization. Related to the former state, this could be called left-handed polarization.

However, seen from the positive pole, both polarities (rotational senses) always appear right-handed, i.e. identical. Therefore, one could believe that a spatial rotation of the left constellation by 180° leads to the right constellation and both situations are physically identical. This apparent invariance seems to reinforce the assumption (later made by Maxwell) that the electric current has a preferred direction and that the polarization of the induced electro-magnetic field is always identical (e.g., a right-handed one if we reference the system from the positive pole). But this could also be a fallacy: since the electric polarity with respect to the conductor is not changed at all when the conductor is spatially rotated by 180°, the two situations cannot be physically identical, although they appear identical. This consideration leads me to suspect that reversing the electric polarity produces a mirror-symmetrically polarized magnetic field, which I therefore mark with a white arrowhead.

What is certain, however, is that the magnetic force always acts transverse to the longitudinal direction of the wire and can "point" in one of two possible directions, depending on the electrical polarity. Thus, an electrically induced magnetic field behaves almost like a cylindrical magnetic monopole stretched in length. The magnetic force is strongest at the surface of the conductor and decreases outward quadratically with distance (just like gravity).

If the battery is clamped the other way round, i.e. if the negative and positive pole are reversed, the compass needle rotates in exactly the opposite direction and settles again transversely to the conductor (Fig.15, right). Thus, the polarized magnetic field points – if it 'points' at all – in one of two possible, exactly opposite directions, which depend on the electrical polarity of the conductor.

Don't polarized light rays behave similarly? Indeed, they show similar magnetic properties, but this was not known in 1820. A current-carrying conductor apparently behaves almost like a polarized light ray. Along the wire, the magnetic field hose has a transverse polarity (like a polarized light ray), but no sources of magnetic force can be identified, i.e., no magnetic poles. Very strange.

In this way, OERSTED discovered the phenomenon of *electro-magnetism* in 1820. The strange relationship between current electricity and magnetism was startling, caused a great sensation and led to hectic research activities throughout Europe. Especially strange was that this kind of magnetic action contradicted the force model of NEWTON and COULOMB. Why do electric fluida or charges, which are supposed to move along the conductor, deflect the needle exactly transversely and not longitudinally? Where does this transverse magnetic force come from? According to NEWTON, all forces, including gravitational, electric and magnetic forces, should act in a direct line. But here the magnetic force acted exactly perpendicular to the electric force and – apparently – circular. This was puzzling. In 1815 OERSTED had already surmised that light and heat are basically the same thing and are based on interactions of opposite electric charges. The discovery that electric currents flowing longitudinally make transverse and polarized magnetic forces now led him to suspect that the polarization of light might also be electromagnetic in nature. He also found experimentally that that the magnetic force penetrates glass and wood and acts only on magnetizable iron. From OERSTED also came the idea to interpret the polarization of the magnetic field as a circular motion (which seemed to be a spiral motion until AMPERE's clarification). However, the erroneous hypothesis of a spiral motion of electric charges gave him already the idea that we might be dealing with *enantiomorphic states of matter* and *mirror symmetric forms of motion*:

All the actions [...] on the north pole of the needle are easily understood if one assumes that the negative electric force or matter passes through a right-handed spiral and repels the north pole, but does not act on the south pole; and likewise all actions on the south pole if we attribute to the positive electric force or matter a movement in the opposite direction [...]". I shall merely add [...] that heat and light are electric conflicts. From the new added observations it can be concluded that the movement in circles also occurs in these effects, which, I believe, can contribute much to the elucidation of those facts which are called the polarity of light." [14]

In the original, however, OERSTED's account is not as clear as I have presented it here. For one thing, he does not speak of magnetic *fields* (although he describes the same thing) nor of electric polarization, but only of an electric 'conflict' (disturbance) that forms around the conductor. He tries to explain the transverse leveling of the magnetic compass needle directly by interactions with opposite electric charges (at this point I have to think of our modern *electron-positron* conception, which describes a spatial-ontological mirror symmetry, i.e. an enantiomorphic property, if one considers both as parts of a larger entity). Translating the enantiomorphic character of opposite electric charges into the field language, the electric current may represent a branched electric field state. If this is correct, we are dealing with a dielectric polarization process. The essence of OERSTED's statement then is that these two "branches" each interact with the magnetically oppositely polarized tips of the compass needle, which rotates it until the forces are balanced. Whether this explanation is sufficient remains unclear at this point.

Ampere's electrodynamics: The bidirectional electric current

Only few months later, in September 1820, ARAGO gave a experimental demonstration of OERSTED's discovery at the French Academy of Sciences in Paris. This extraordinarily impressive demonstration inspired AMPERE to conduct his own research into the nature of electric currents and magnetic actions. Carefully experiments in

14   Oersted, Juli 21, 1820

his own laboratory led him to a mathematical- physical theory of electro-magnetism, which he called *electrodynamics*.

At first he repeated OERSTED's experiments and found that the deflection of the compass needle was not 45° but exactly 90° when the Earth's magnetic field was compensated with a suitable device – invented by himself and called an *astatic compass*. To explain the magnetic action produced by electric currents, he postulated an *electrodynamic force*, which is the polarized electric force that turns the magnetic needle. In other words, AMPERE doubted the existence of *magnetic fluida* or *fields* and wanted to attribute magnetism exclusively to electric forces and direct remote actions as in the theories of COULOMB and NEWTON. Obviously, dipolar properties played a decisive role here as well, which he tried to integrate into Newtonian theory. AMPERE assumed that the current makes an electrodynamic force with a certain direction of rotation around the conductor. In this way, the polarity of the magnetic fluida or field, which in reality need have nothing at all to do with rotational motion, is interpreted as an induced circular *electric current* flowing either clockwise or counterclockwise around the conductor. This is another switch of cognition: From now on, the idea prevailed that the properties of the polarized magnetic field – or of the polarized state of an electrically excited space – could have something to do with directions of motion, rotational movements and real flows in reality. Which of the two versions of the electrodynamic force acts on the magnetic needle depends on the electric polarity of the wire (as the polarity of the magnetic field). AMPERE's notion of an electrodynamic force thus corresponds to the notion of a polarized magnetic field, but is now associated with a circular electric current, with real motion. He assumed that such circular currents also exist in permanent magnets and ordinary matter, called AMPERE's molecular currents, and generate what we call magnetic fields since FARADAY. The interaction between wire and magnetic needle is then explained as the mutual action of two electric 'current elements' (between two pieces of current carrying wires or circular currents) which instantaneously exert electric forces on each other, acting either repulsively or attractively.

AMPERE illustrated this concept by explaining the magnetic field of the Earth as follows: within or around the Earth, a circular electric current flows in an east-west direction and generates transverse

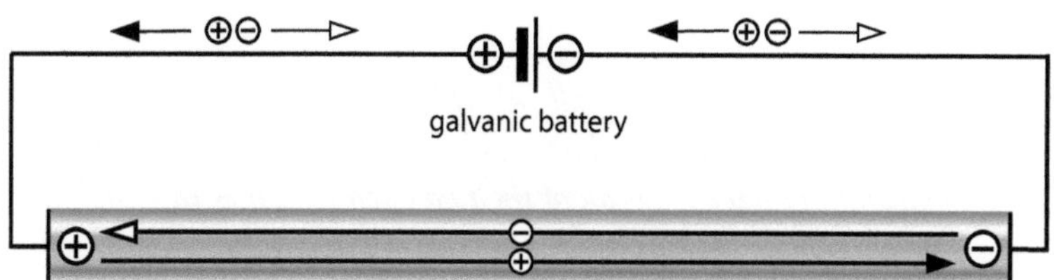

Figure 16 Ampere's bidirectional current (e⁺+ e⁻)

The lower segment of the wire in the circuit is shown as an enlarged cylinder (gray) to illustrate the similarity between electric current in metallic conductors and dielectric polarization in insulator materials (figures on the next pages). According to Ampere, opposite electric charges move in opposite directions in the metal wire. Ampere defines a preferred direction of electric current from plus to minus (black arrows) only for didactic reasons, to be able to unambiguously define the sense of the electrodynamic force (aka polarity of the magnetic field), which depends on the electric polarity of the conductor. This reference definition is known as the technical definition of the electric current. In modern physics, however, the direction of current is arbitrarily defined from minus to plus, because since Maxwell it has been assumed that the electric current consists only of negative electric charges.

electrodynamic forces polarized in north-south direction. Similarly, a molecular electric current should orbit transversely around the dipole axis in a permanent magnet. AMPERE proved this idea by two current- carrying solenoids (copper spirals), which behaved indeed as if they were permanent magnets.

Sometimes it is claimed that AMPERE proved that the electric current has a certain direction. However, this is a myth. In AMPERE's theory, the electric current has no preferred direction, quite the contrary: It is a *bidirectional current*. However, one learns this only if one reads his original papers.[15] In AMPERE's electrodynamics, negative and positive electric charges move past each other in opposite directions, representing a double current and bidirectional motion (Fig.16). However, correct is that he defined a *reference direction* of the electric current from the positive pole to the negative pole in the external circuit (outside the battery), but only for didactic purposes, not for physical reasons. AMPERE explicitly pointed out that he defined the direction of electric current as the motion of positive charges from plus to minus only for one reason: In order to be able to unambiguously define the rotational sense of the electrodynamic force *relative* to the electrical polarity of the conductor, i.e., to an arbitrarily introduced direction of the electric current, and to avoid the constant repetition of the fact that the electric current has bidirectional character:

15 which A.K.T Assis has excellently prepared and translated into English (see footnote next page)

„For the sake of simplicity I shall call this state of the electricity in a series of electromotive and conducting bodies electric current; and since I shall continually have to speak of the two opposite senses in which the two electricities move, I shall invariably imply 'positive electricity' by the words 'sense of the electric current' to avoid unnecessary repetition."[16]

AMPERE's definition of the direction of the current is still considered the 'technical definition', while in physics, paradoxically, a reverse direction of electric current from minus to plus is assumed, since many physicists would like to understand the electric current (not entirely unbiased) as the motion of particle-like electrons.

To determine the relationship between electric and electrodynamic (aka magnetic) polarity, AMPERE positioned his precision galvanometer (basically a highly sensitive compass) at various points on the circuit and battery. He notes that the magnetic needle on the two wires coming from the battery poles point in the same direction, while it points in the opposite direction around the battery. At first glance – if one takes the didactic current definition of AMPERE too literally – this seems to indicate that the electric current in the battery flows in the opposite direction, from negative to positive. But since electric currents are supposed to be bidirectional, this observation actually only says that the electrodynamic force (aka magnetic force) around the battery shows the opposite sense. Translated into the field model, this means that the polarity of the magnetic field around the battery is opposite to the polarity of the magnetic field that forms around the conductors coming from the battery poles. Thus the circuit as a whole seems to represent a magnetic dipole.

AMPERE also found out that an electric current in a wire bent into a circular or rectangular shape produced a magnetic torque that acted on the entire circuit. This apparently had to do with the polarized electrodynamic force (the polarized magnetic field) around the conductor: when the wire changes direction, the sense or direction of the electrodynamic force also changes. As a result, the *relative* direction of the electrodynamic (magnetic) force appears opposite on the opposite sides of the circuit. This phenomenon was one of the rea-

16 Ampere 1820, quoted from: Assis & Chaib: "Ampere's Electrodynamics", Keys Inc., Montreal 2015

sons for FARADAY to develop the idea of a magnetic field which conditions the space between current-carrying conductors (even if they belong to the same conducting loop). And this was also the reason for MAXWELL's conclusion that the magnetic field around the wire must be polarized. If one illustrates this polarization like MAXWELL with a rotational direction of the magnetic field, the torque in the electric circuit can be explained approximately in this way:

If you mark the circumference of a hose at regular intervals with arrows that always point in the same direction and bend the hose in a rectangular or circular shape, you will find that the arrows at the opposite segments will point in the opposite direction – clockwise and counter- clockwise. Surprisingly, this bending has a real physical effect: In an external magnetic field, the opposite segments of a current-carrying wire loop are subjected to opposite forces. One side is magnetically attracted and the other is repelled. This creates a torque that attempts to rotate the current loop. This was the actual discovery that would later lead to the design of electric motors. This strengthened many physicists of that time – but not AMPERE – in the assumption that an aether gas must exist, whose electrically induced deformations produce magnetic dipole states, on which the magnetic field of the Earth (or an external magnetic field) acts differently and thereby causes a kind of 'molecular' rotation of the magnetized aether gas. So the question arose what really happens in or around a current-carrying wire, what an electric current actually is, and what electrodynamic forces – or polarized magnetic fields – distinguishes from ordinary magnetism, if there is a difference at all.

In order to be able to describe the relationships between electric and magnetic polarity in a circuit, AMPERE introduced a *fictitious observer*. His only task is to provide a *spatial reference system* and *relative directions*, since terms like left/ right, up/ down, forward/ backward or clockwise/ counterclockwise have no physical meaning. Space as an isotropic volume has no preferred directions. Spatial directions are relative assignments whose meaning depends on the reference system which is arbitrarily determined locally by humans. Thus, it is important to understand that AMPERE only wanted to illustrate the relative dependence between the polarity of an electric current and the polarity of the electrodynamic (or magnetic force) generated by it. Physically, this has nothing at all to do with spatial directions, but rather with physical structures that embody a left-right mirror symmetry. This has a physical meaning because it characterizes an ontological nature, namely a mirror-symmetric structure.

If this is not clear, the arrow illustration of polarization can easily be confused with spatial directions that exist in reality, leading to fictitious ideas, e.g. of a preferred direction of motion of electric charges, circling electrical currents, or magnetic field vortices. This is reminiscent of FRESNEL's planes of polarization, which appear to exist in space in a well-defined way, but according to quantum physics cannot have a predefined orientation in space before an interaction has taken place (which is called a "measurement"). Ambiguities of this kind occur both in MAXWELL's electromagnetic field theory and in quantum mechanics. But if we are dealing with dipolarized fields in reality, there is no reason to assume that they rotate or move in the sense of mechanics, quite the contrary: they are probably stationary or pseudo-static dipole fields. Around current-carrying conductors, they appear as tubular magnetic fields with magnetic polarity but without local magnetic poles, what could be understood as polarized magnetic fields branches. So, as in the case of AVOGADRO's molecule theory, we can say that the roots of the problems that cause us such a big headache in understanding quantum physics are already two hundred years old.

That electric and magnetic polarization processes are the essence of AMPERE's electrodynamics is shown by an experiment he devised to define the intensity of an electric current indirectly by the strength of the mutually interacting electrodynamic (magnetic) forces. The experimental setup consists of two vertical parallel wires through which an electric current flows. The current intensity is the same in both. When the wires are electrically polarized in the same sense, expressed as 'both currents have the same direction', or 'both currents flow in parallel', they attract each other magnetically (electrodynamically, in AMPERE's terms). When the wires are electrically oppositely polarized, expressed as 'the currents have opposite directions', or 'the currents flow anti-parallel', they *repel* each other

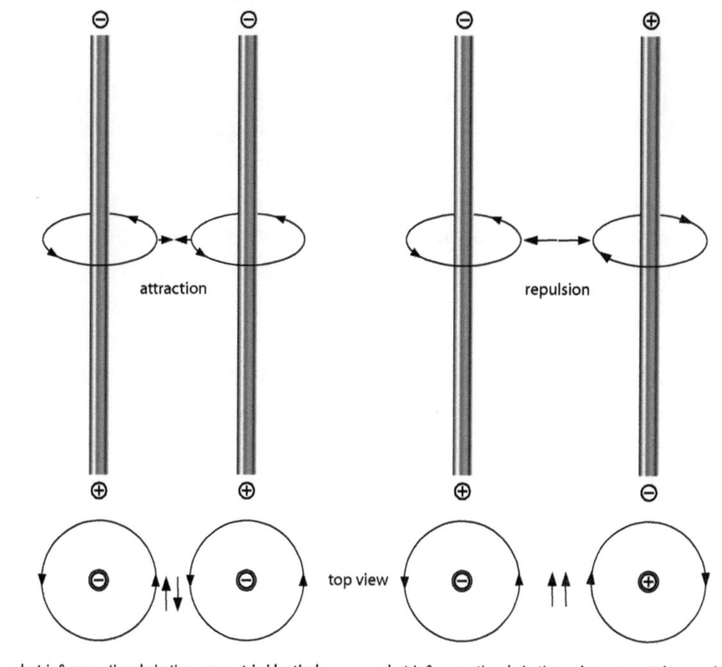

Figure 17 Attraction and repulsion of two current-carrying conductors
Conductors with the same magnetic polarization attract each other (left), conductors with opposite magnetic polarization repel each other (right). This seems to contradict the magnetic theory, according to which equal magnetic polarities repel and opposite ones attract.

magnetically. This is irritating because equal polarities should repel and opposite polarities should attract – as described by COULOMB's law. In fact, however, it is the other way around: Although both magnetic fields have the same polarity, since the electric polarity is the same, the magnetic fields attract each other (Fig. 18, left). This reminds us of the discussion in the OERSTED section (see Fig. 15, p. 111). That was later to confuse MAXWELL as well, when he tried to model the rotation of neighboring aether cells (we will come back to this later in the section about his interpretation). This contradiction can apparently only be resolved if one assumes that the transversal field line around the wire represents either a circular electric current or a magnetic eddy or flux. This is the reason why magnetic polarization is modeled with a circular motion: Where the two field lines of the two conductors touch locally, the magnetic polarizations and flux directions then point in opposite directions, although the electric and magnetic polarizations are identical (Fig. 18 left). The additional assumption of a directed motion in the form of a magnetic flux or vortex was thus introduced to be able to explain why two electrically induced magnetic fields with the same polarization attract each other instead of repelling each other. With this subtle additional assumption, the magnetic theory appears coherent again:

Local opposite magnetic polarities attract and equal ones repel. However, this explanation presupposes that the circular motion of magnetic fields (or electric currents) really exists. If this movement does not exist, we can only deal with static, i.e. standing magnetic fields. Then there must be another explanation, which could have to do with the induction process that FARADAY discovered only much later, or simply with the fact that magnetism is not yet fully understood.

The magnetic attraction or repulsion force decreases quadratically with the distance between the wires and is directly proportional to the intensity of the electric currents. From AMPERE also came the new concept that the electricity flows inside the conductor and can be modeled like the flow of a liquid in a tube, except that dynamic electricity consists of two opposite currents. If one imagines electric charges as volume units passing a cross section of the conductor, the current intensity must be proportional to the number of positive (and negative) charges passing per time unit multiplied by their speed. However, it was not clear whether these electric charges or volume elements really exist, how they are set in motion, how fast they move and how this speed can be defined. To explain how an electric current is created, AMPERE postulated an *electromotive force* that accelerates opposite electric fluids or charges in opposite directions, setting the two currents in motion. The electromotive force (abbreviated: emf) thus describes an acceleration not yet known in Newtonian theory, and, at the same time, a process of charge separation that creates a dynamically stable physical state:

„The currents accelerate until the electromotive force is balanced by the inertia of the electric fluids and the resistance of the conductors, whereupon they continue indefinitely at a constant speed so long as this force conserves the same intensity, but they cease instantly whenever the circuit is interrupted.“ (Ampere, 1820)

And exactly this is the next switch of cognition: As in the case of the molecule hypothesis and the wave theory, holistic division and field branching processes may be discovered also in the electromagnetic theory. They are hidden behind the concept of "electromotive" force and represent polarizing branching processes. This means that the opposite electric charges are not present a priori, but are created by branching processes. This structure formation process corresponds to an acceleration – which is not mechanical in nature –, provides a new concept of kinetics and leads to stable dynamic or pseudo-static branching states, which are in accordance with the actio-reactio principle, the principle of energy conservation and the symmetry-principle of relative motion. In short, the branching principle could also be applied in electromagnetic theory, explain what polarization is, and be unified with AMPERE's bidirectional current within a field model.

Summary:

The discovery of OERSTED and the research of AMPERE led to the hypothesis that the electrically induced magnetic polarization could represent a rotational motion of the electrodynamic force (or magnetic field), which occurs in opposite directions depending on the electrical polarization. This characterizes the magnetic polarization as the direction of a magnetic vortex. But was that really proof yet?

If one limits his consideration only to one section of the conductor, one can easily overlook that the circuit including the battery forms a holistic physical system consisting of two opposite magnetic polarizations (one around the wire, the other around the battery), representing a magnetic dipole. Although AMPERE questioned magnetic fields as physical elements of reality, he undoubtedly measured magnetic influences with his astatic compass and special developed galvanometer (which is why I take the liberty of speaking of magnetic fields, which were introduced only much later by FARADAY). AMPERE, however, was looking for a purely electrical explanation: As a true follower of NEWTON, he preferred the explanation of magnetic actions as direct instantaneous actions. A mediating medium for the forces was not required, neither a field nor a gas. This was a reasonable point of view, because it is impossible to prove the existence of fields directly – even today. The only thing we can detect when we measure something is an *effective* (i.e. structure-changing) interaction with or in our measuring devices. Although this is actually a physical banality, this fact should lead to deep confusions in quantum physics one hundred years later (we will come back to this several times).

It can be said without doubt that the interpretation of the discoveries of OERSTED and AMPERE has a key significance for the understanding of the true constitution of nature. AMPERE's conception of circular electric currents and molecular currents is still in use today to explain magnetism and is even considered proven. It is supported today by quantum spin models which use the same idea to explain the magnetic properties of electrons, atoms, molecules and all other 'elementary' particles with rotations about their axis. But this picture cannot be the right one, because it maintains the existence of particles and suggests the idea of a mechanical motion. Both ideas are not only not provable, but are clearly refuted by quantum physical double-slit experiments. So, both ideas must be wrong, at least concerning the fundamental nature of matter and light. So it could well be that the idea of polarized stationary magnetic fields might be the more appropriate one.

AMPERE pursued an electrical interpretation of magnetism because he believed that things that can interact must have the same nature. If magnetism were anything other than electricity, electric currents could not interact with magnets. But if this philosophical assumption is legitimate, the opposite could also be true: the bidirectional electric current could just as well be a manifestation of magnetism. If this is true, magnetic forces should also be able to make charge separations, i.e. dielectric polarizations. This is exactly what FARADAY discovered ten years later: the magneto-electric induction.
As far as I can see, the concept of branching processes fits quite well with the conception of bidirectional current and the electromotive force. If mechanical concepts, especially the indivisibility assumption and thus the particle model, are wrong and have to be replaced by holistic division and branching processes, then it becomes clear that a structural bifurcation movement can no longer be understood

and modeled as a unidirectional motion. If AMPERE's bidirectional motion is interpreted as a holistic division process rather than a mechanical motion, the electric current turns out to be nothing more than a dielectric polarizing branching process. Why the conception of bidirectional current was abandoned remains to be clarified. [17]

Our attempt to explain the nature of electric current by dipolarizing field branching processes is motivated by yet another problem that has received too little attention: As we will see in the next sections, there are still contradictions in the explanation of electromagnetic induction, which, among other things, go back to the hypothesis of circular electric currents or magnetic vortices around conductors. The resolution of the problem begins with the insight that magneto-electric induction does indeed prove that magnets produce electric charge separations, i.e. bidirectional movements and dielectric polarizations. But this is again my own interpretation, which differs from the textbook view, but agrees just as well with the experimental facts. So this could also be a discovery. To understand how this new model works, we need to develop a pictorial idea of what the field concept means and why it seems so useful and real, even if the existence of fields (like the aether) cannot be proved directly. Then we will understand how the interpretation of the nature of electric current and electromagnetic induction relates to the concept of the speed of light. This will shed new light on special relativity and the contradiction between entangled (branched and dipolarized) quantum states and the unidirectional definition of the speed of light.

17   We will discuss that in the section 'Maxwell's field theory'. Maxwell adopted Ampere's mathematical definition of electric current but not the concept of a bidirectional current. Although this question could not be decided experimentally, at the end of the 19th century the assumption that an electric current consists of only negative charges and flows in only one direction became tacitly accepted.

Faraday's field concept and the magnetic nature of light

The English chemist and physicist MICHAEL FARADAY (1791-1867) also repeated the experiments of OERSTED and AMPERE in 1821 and developed immediately a new idea: If electric currents induce magnetic forces, it should also be possible that magnetic forces induce electric currents. Although he had not yet succeeded in proving this, he was able to realize a first electromagnetically rotating device, later known as homopolar motor. But it was not until ten years later, in 1831, that he discovered magneto-electric induction and was thus able to prove that magnetism can indeed be converted into electricity. So they two had to be close relatives, if not two sides of the same coin.

FARADAY, originally a journeyman bookbinder, was so enthusiastic about natural science that in his spare time he preferred to attend physics lectures and carefully studied all the books he had to bind. Due to his curiosity, imagination, self-motivation and autodidactic abilities, he managed to become an assistant to the most respected British scientist in 1813, without official scientific training. Within only fifteen years, FARADAY himself developed into the most famous physicist, experimenter and natural philosopher of the Royal Society. He was a truly extraordinary man who rigorously refused to be ennobled by the King for his scientific merits. His public London Friday evening lectures on physics, chemistry and natural philosophy, which were entirely devoid of mathematics, were legendary at the time.

FARADAY realized the liquefaction of gases under pressure, studied metal alloys and heavy glasses, produced new chemical compounds and developed the idea of magnetic and electric 'force lines' which led to the field model in 1846. While investigating voltaic batteries, he discovered that the integral and compound molecules involved dissociate in dilute acid or salty solutions (hence a division process), thereby acquiring an electric charge, migrating to the opposite electric poles, and generating an electric voltage between the poles. The amount of material deposited at the electrodes corresponds to the amounts of electrical charge transported. Molecular decomposition and charge transport take place in both directions. In this way he discovered the crucial role of electricity in the formation, decomposition and bonding of molecules, which made him the founder of electrochemistry.

FARADAY concluded from the experiments that something real must exist in the space between the bodies, which transmits electric and magnetic effects. He did not believe that these are electric or magnetic fluida (gaseous substances), but pure force fields that represent a form of existence of their own. He concluded this from electrolysis experiments, in which he had studied the functioning of galvanic batteries and the electrical decomposition of the integral molecules that made up the electrodes of the battery (the positive and negative poles), and from experiments with electrical discharge tubes containing extremely diluted gases that came quite close to a vacuum. All these experiments indicated that electric and magnetic forces are independent of the respective nature of liquids and gases and can also act in a vacuum. That is, electricity is always of the same nature regardless of its origin. It makes no difference whether the charge separation process involved is of chemical (electrolytic), electrostatic (friction) or electrodynamic (magnetic) nature. He also discovered that there must exist a smallest unit of electricity, which he called *ionic charge.* Ions are integral molecules or atoms that are no longer electrically neutral because they have either given up or taken up such an electrical charge unit. Sixty years later, physicists began to interpret this smallest unit of electricity as a negative charged particle and call it 'electron'.

If one plays around with two permanent magnets, one can directly feel that there must be something invisible around magnets, which sometimes behaves like a thick, repulsive pad and sometimes like a suction cup. That is exactly what FARADAY called a magnetic field. Equal magnetic poles repel each other and opposite magnetic poles attract each other. To make the shape of magnetic fields visible he used fine iron filings. In the magnetic field, they arrange themselves along 'force lines' by a multitude of *local magnetic polarization* processes. That means that the magnetic field simultaneously magnetizes all the iron filings in its vicinity (a process that bears holistic features). A new, local magnetic field with a north and a south pole is created in each of the iron filings. They behave like tiny magnetic dipoles and align parallel to the field lines, revealing the shape of a force field, the magnetic field. In this way, FARADAY was able to show how a magnetic field conditions the environment and exerts forces on suitably receptive materials.

However, FARADAY's field concept went far beyond what AMPERE could ever accept, which led to some discussions (they corresponded by letter). In AMPERE's concept, magnetic fields are not needed, for magnetic actions are explained exclusively by electrodynamic forces of attraction and repulsion which act instantaneously over arbitrary distances. But according to FARADAY's concept, the electrodynamic force is nothing else than a magnetic force. In FARADAY's model, a current-carrying conductor is simply *magnetized* and surrounded by a magnetic field hose, represented by circular field lines with a certain magnetic "direction" which depends on the electrical polarity of the conductor. This was his interpretation of OERSTED's discovery. But what seemed to be not clear at this time is that the magnetic hose around a current-carrying wire is a homopolar magnetic field, thus a mono-polarized magnetic field.

In 1831, FARADAY discovered magneto-electric induction after realizing that electricity and magnetism can indeed be transformed into each other, but only if something *moves* or *changes*. A magnet must move relative to a wire, or a wire relative to a magnet, to induce an electric current in the conductor (Fig. 18-20). Thus the mechanical induction effect depends mainly on the change in distance, which changes the magnetic field strength locally at the point of action. This is the so-called *symmetry of motion* of electrodynamics, which was later to become so famous as the principle of relativity. Another possibility to get the same induction effect is by using an electromagnet: When switched on, the electric current creates a magnetic field around the coil. Just like an approaching permanent magnet, this suddenly appearing – or changing – magnetic field induces an electric current pulse in a nearby metal rod or wire. The brief electric current pulse creates an electrical polarity, detectable as a positive electric charge at one end and a negative electric charge at the other end of the rod or wire, i.e., positive and negative poles. If one

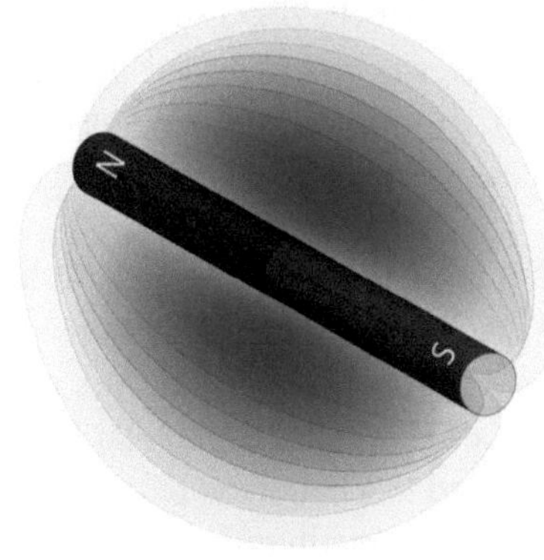

Figure 18 The magnetic field

Doesn't it look like an apple? Of course, a magnetic field is invisible. This picture is only to give us a spatial idea of the shape of the magnetic field. It should be pointed out that the magnetic field has no real boundary, but gets weaker and weaker towards the outside. Faraday made the shape of the magnetic fields visible with iron filings and illustrated their alignment with imaginary field lines.

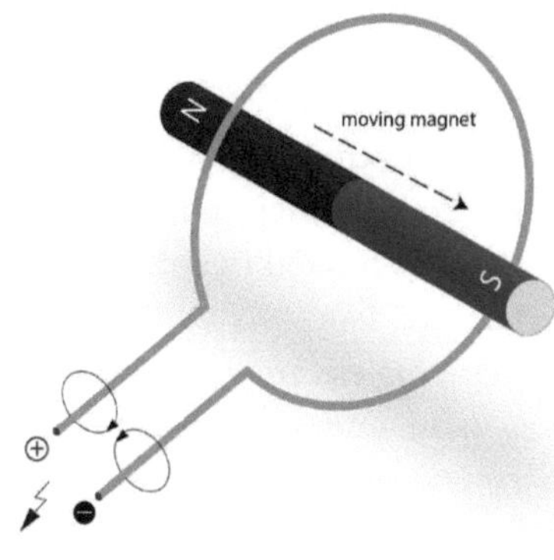

Figure 19 Electromagnetic induction

The motion of the magnet relative to the wire loop or vice versa induces an electric current in or along the wire. One also speaks of an electric field that generates a voltage. Opposite electric poles are created at the ends of the wire, i.e., the wire gets dielectrically polarized *and* magnetized. When the ends of the wire are close together and the voltage is strong enough, it can produce a sparkover that causes charge equalization (neutralization), resulting in the emission of light. Any change in the distance between the magnet and the wire causes a change in the magnetic field strength at the location of the conductor and makes a short current pulse. If an electromagnet is used, the current pulse is generated without a change in distance. The decisive factor is therefore the change in magnetic field strength at the location of the conductor.

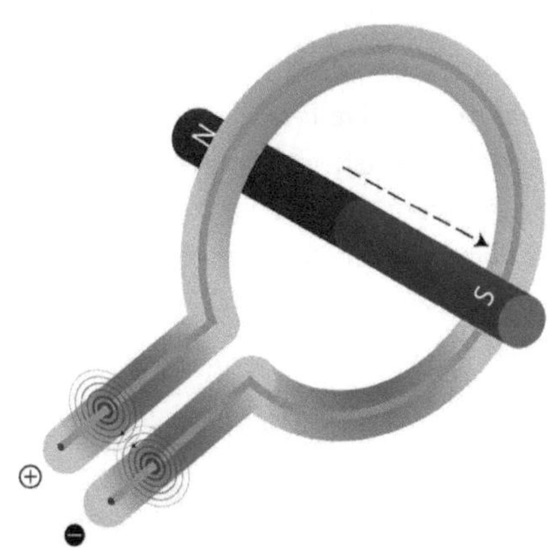

Figure 20 Polarized magnetic field hose

The induced electric current pulse makes a monopolar magnetic field hose around the conductor. The shape and strength of the magnetic field can be illustrated with concentric field lines, which represent imaginary cross-sections through the magnetic field and symbolize cylindrical surfaces of equal magnetic field strength. The magnetic polarization of the tube can be indicated by arrows (as in Figure 20).

turns the permanent magnet around so that the opposite magnetic pole approaches the wire, or if one reverses the electrical polarity of the electromagnet, the electric current pulse in the wire has the opposite polarity: minus and plus pole are reversed.

FARADAY recognized early that electromagnetic induction could be used to design electric motors and generators. AMPERE had already shown that opposing magnetic forces act on the opposite sides of a current-carrying wire if it is bent into a right-angled or circular loop. If the current-carrying wire loop is placed in a magnetic field, opposite forces act on the magnetic polarities, attracting on one side and repelling on the other, causing either the wire loop or the magnet to rotate, depending on which part is fixed and thus becomes the stator (action = reactio). This is called the *magnetic momentum*, which means a torque. The maximum rotation is 90 degrees, because then a balance of forces is achieved. However, if the polarity of the electric current is changed at that moment, a continuous rotation can be realized. This is the principle of the electric motor. The reversal is an electric generator: when the wire loop is rotated in a magnetic field, an electric current is induced in the coil. FARADAY's induction principle is summarized as follows: a changing electric field induces a magnetic field (*electro-magnetic induction*), and a changing magnetic field induces an electric field (*magneto-electric induction*). The only question is whether, when we talk about the "electric current", we are dealing in reality with a unidirectional motion in the sense of mechanics or with a bidirectional movement in the sense of a field branching and polarization process. In contrast to the usual interpretation by MAXWELL, who assumed that the electric current flows in only one direction and consists only of negative charges, electromagentic induction can be interpreted differently if one questions the underlying assumptions. As it turns out, one of the assumptions can be made more precise and the other can be replaced by its opposite:

1) The electrically induced magnetic field is mono-polarized, or in short, a polarized magnetic field. It has a 'preferred direction', but magnetic poles cannot be identified. This is a specification justified by the experiments of OERSTED and AMPERE.

2) The moving permanent magnet makes an electric charge separation in or around the conductor, i.e. an electromotive force, which can be interpreted as a bidirectional branching motion and holistic division process, which causes a dielectric polarization. This is the so-called 'electric field' that makes the current pulse.

This is the new idea. So, if we unite AMPERE's concept of bidirectional electric current with FARADAY's field concept and the hypothesis of holistic division and branching processes (as suggested by the double-slit experiments of quantum physics), electro-magnetic induction could be understood as a field branching process generating a dipolar, enantiomorphic electric field: The approaching magnet induces an electric current pulse/ voltage peak in or around the conductor, consisting of two opposite half-charges moving in opposite directions, which can be detected as positive and negative poles at the ends of the wire (this is how the *voltage* is measured). This picture is strikingly similar to the branching of a light beam and the resulting magnetic dipolarity of the two partial rays in optical experiments. The only difference is that the branched field state is not as nicely visible as with light rays, that there is an electric rather than a magnetic dipolarity involved, and that the electric field branching process seems to be tied to a piece of metal. In such a model, charge separation, bidirectional current pulse and field branching are the same. Thus, magneto-electric induction makes an *enantiomorphic* electric field, a holistic field which is oppositely structured in itself. This is a structural movement, a branching process. In this way the field concept could be united with the bidirectional electric current concept. In this model, the electric current is not a 'current' at all, but a field branching process which makes a dielectric polarization (just like in insulator materials), which can be measured as an electric current pulse. This explanation is so obvious that one wonders what is so complicated about FARADAY induction and the symmetry of motion. Obviously, the problem is that Newtonian mechanics and field theory do not know branching processes at all.

Now another question arises: Can we reverse AMPERE's philosophy? Can we explain electric phenomena exclusively by magnetic fields? Let's try it: the magnet approaches, with one pole facing forward, a metal rod or wire lying across it. It then acts like an inhomogeneous magnetic field – almost like a magnetic monopole – and induces a magnetic counter-field at the conductor, which branches out. This creates opposite 'magnetic poles' at the ends of the metal rod, which appear to us as opposite electric charges. So in this picture, the electric current pulse is just the manifestation of a *branching* magnetic field (on a metal rod).

This case can be compared to the branching of an already polarized light beam, resulting in magnetically oppositely polarized partial waves or rays. The same process can be discovered in the division of an integral molecule (aka atom) in the the double-slit experiment, as soon as one takes the division process physically seriously: the branching generates magnetically opposite polarized partial waves, which expand further. This is an expanding molecular structure, which now has a magnetic dipole character. If this is true, one could say that a holistic division processes of magnetic fields create magnetic dipole structures and thus the complex structure of matter.

But how can such a magnetic division processes explain the attraction and repulsion of electric charges that seem so significant in our atom models? In other words: how are electrostatics and magnetism related? We will return to this aspect, which could also be related to the phenomenon of *diamagnetism*, in a moment, and again later when we discuss the STERN-GERLACH experiment with silver atoms. But for now, let's stick with the most obvious idea, that the relative motion between magnet and conductor generates an bidirectional electric current pulse:

Case A: The magnet is moved, the metal rod is at rest
The magnet approaches – with a magnetic pole ahead – a transverse copper or silver rod. From the view of the metal rod, this represents an inhomogeneous magnetic field, appearing almost as a magnetic monopole. This polar magnetic field induces the branching of another field in or around the conductor, which we call *electric field* after its formation. We recognize in this induction process the formation of an enantiomorphic electric field, an electric charge separation, a dielectric polarization, and – in contrast to MAXWELL – a bidirectional current pulse leading to a short voltage spike (Fig.21).

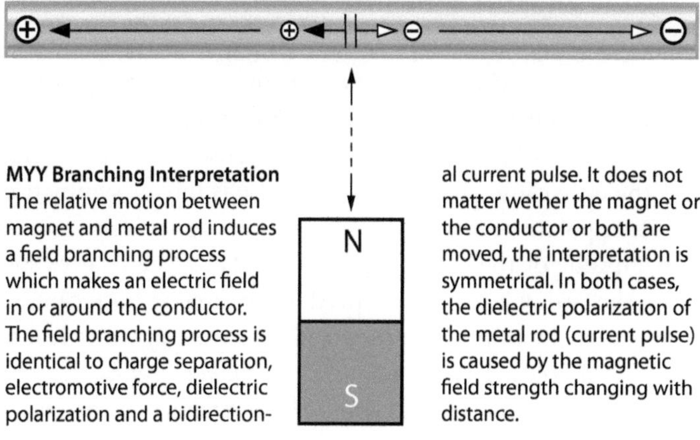

MYY Branching Interpretation
The relative motion between magnet and metal rod induces a field branching process which makes an electric field in or around the conductor. The field branching process is identical to charge separation, electromotive force, dielectric polarization and a bidirection-al current pulse. It does not matter wether the magnet or the conductor or both are moved, the interpretation is symmetrical. In both cases, the dielectric polarization of the metal rod (current pulse) is caused by the magnetic field strength changing with distance.

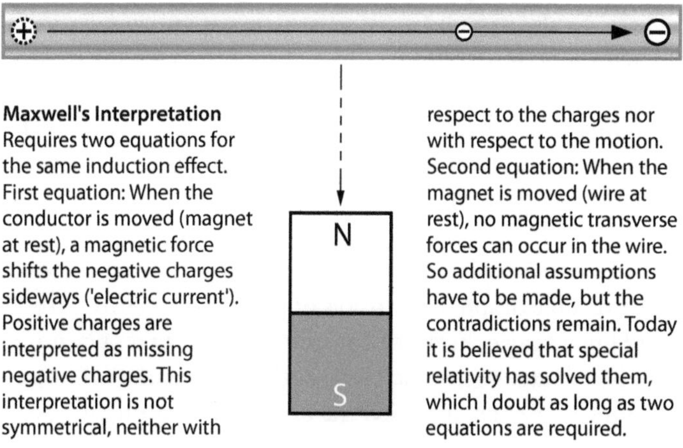

Maxwell's Interpretation
Requires two equations for the same induction effect. First equation: When the conductor is moved (magnet at rest), a magnetic force shifts the negative charges sideways ('electric current'). Positive charges are interpreted as missing negative charges. This interpretation is not symmetrical, neither with respect to the charges nor with respect to the motion. Second equation: When the magnet is moved (wire at rest), no magnetic transverse forces can occur in the wire. So additional assumptions have to be made, but the contradictions remain. Today it is believed that special relativity has solved them, which I doubt as long as two equations are required.

Figure 21 Electromagnetic induction
Top: My branching interpretation. Bottom: Maxwell's interpretation

Case B: The metal rod is moved, the magnet is at rest

The transverse metal rod approaches the magnetic pole and senses an inhomogeneous magnetic field. This polar appearing magnetic field induces the branching of another field in or around the moved metal rod, which we call *electric field* after its formation. We recognize in this induction process the formation of an enantiomorphic electric field, an electric charge separation, a dielectric polarization and a bidirectional current pulse leading to a short voltage spike.

In other words, in our interpretation, the explanation of induction is exactly the same whether the magnet, the conductor, or both are moved (this is not the case in the interpretation of Maxwell; we will return to this later). This is simply because the polar magnetic field is the only cause and the current pulse is mirror-symmetrical. The electric polarity remains the same if the magnet is not reversed. The magnitude of the voltage spike depends on the relative speed between the two objects. If this speed is the same in both cases, the voltage is also the same. The motion is therefore symmetrical and the induction effect depends only on the *change in distance*, which causes a change in the local magnetic field strength at the location of the conductor, regardless of whether the magnet or the conductor is moved. This is what is meant by *relative motion* in mechanics.

However, the change of the magnetic field strength at the location of the conductor can also be caused by a stationary electromagnet. Then the term *relative motion* can no longer be interpreted in the sense of Newtonian mechanics as a *change of location*. So there must still be a non-mechanical relativistic interpretation – and this can only be a structural change. That is also a relative movement, even if nothing moves in the sense of mechanics.

Faraday induction also proves that the bifurcation of the electric field is reversible, implying a fusion of the field structure: The electric field branching or electric dipolarisation only takes place as long as there is a change of distance. When the relative motion stops, the reverse process begins, in which the branched structure merges and charges neutralize. This counter movement induces a new cylindrical magnetic field around the metal rod, just like the original electric current pulse, but now with opposite magnetic polarity. Relative to the magnet, however, the new field has then the same polarity, which repels the magnet, or, since actio=reactio: The metal rod will repel itself if the magnet is much heavier or is fixed somehow. In this interpretation, the Faraday induction experiment shows a symmetry of all motions involved, which is exactly what is meant by the term *relativity of motion*. So relativity is not only about the symmetry of mechanical motion (change of location of bodies caused by forces), but also about the symmetry and reversibility of structural movements (branching and fusion).

At this point, however, I must note again that this interpretation of FARADAY induction is not the textbook standard. In the interpretation of MAXWELL, electromagnetic induction is interpreted somewhat differently with a unidirectional motion of exclusively negative charges, which are supposed to represent the electric current in the wire. I think this is exactly the point where the problem begins that seem so intractable later. We will come back to this key problem, which is at the center of contemporary physics, several times.

So far, we have not been able to clarify the question of whether the electric current flows inside or outside the wire. However, FARADAY provided a decisive clue as early as 1845:
He discovered that non-magnetic metals like copper, silver, brass, bismut and almost all non-metallic substances response to external magnetic fields with an opposing magnetic field, which represents a kind of shielding. FARADAY called this originally *dimagnetism* (in analogy to dielectrics), later *diamagnetism*. The latter notion is a bit irritating, because it actually means that non-magnetic materials are permeable to magnetic fields, as OERSTED had already noticed. But how can this be, if the material reacts to an incident magnetic field with an oppositely polarized magnetic field – and thus repels it?[18]

Is this rather a *dimagnetic polarization* analogous to the dielectric polarization of the glass of the Leyden jar? Remember that glass is an electrical insulator but reacts in a similar way to electric fields: Although glass is non-conductive in the usual sense, glass reacts to an external electric field with a dielectric polarization of its internal structure, which obviously represents a *holistic* structure formation. The effect: On the outer sides where the voltage is applied, the glass body forms opposite electric poles and thus counteracts and even neutralizes the external electric field. Where the positive pole of the voltage makes contact the glass surface reacts with a negative pole. On the other side, where the negative pole (electrode) makes con-

18   The shielding effect applies to almost all non-magnetic materials such as water, gases (e.g. helium), wood, plastics, apples and even living creatures, but also to metals such as bismut, gold, copper and silver. The emerging anti-magnetic force can be used to levitate non-magnetic objects with the help of strong magnetic fields. When the experimental prove was given in 1997 by Andrey Gaim, University of Nijmegen, Netherlands, physicists could hardly believe it. There are some quantum physical models, but none is perfect. So, it is still too early for antigravity machines and force shields - first we have to understand quantum physics.

tact, the glass surface forms a positive pole. At the same time, the glass itself is under a certain mechanicals stress. The glass body *as a whole* thus responds to the external field by forming an internal, mirror symmetrical anti-field. This is the nature of dielectric polarization, which makes materials such as glass, ceramics, wood, gases, etc. good electrical insulators.

Traditionally, however, one tries to model dielectric polarization with a particle model in combination with the field concept: It is assumed that negatively and positively charged particles are already present in atoms and molecules, which are only shifted from their rest position by external electric fields. Thus, electric fields are supposed to cause a spatial displacement of already existing charged particles. This explanation is unsatisfactory for two reasons: First, in the normal state, these charges are supposed to be on top of each other or so close to each other that they neutralize and do not exert electrical forces. In other words, this means that their existence in the neutral state is not provable at all. Thus, the particle conception is only an ad hoc hypothesis within the framework of a constructive theory invented to explain dielectric polarization and other processes. Second, the double-slit experiment confirms that the existence of indivisible electric *particles* or *invariant volume elements* is indeed purely fictitious, since the divisibility condition definitely excludes atomistic particle ideas of any kind. Instead, quantum physics challenges us to design a new and more reasonable physical model of the constitution of nature. Thus, if these hypothetical particles cannot exist at all, the phenomenon of dielectric polarization can only be understood in this way: It must be a field branching process which creates opposite electric charges and a mirror symmetric field structure, i.e., an enantiomorphic field (Figure 22). This, however, seems to contradict a fundamental assumption of physics. As RICHARD FEYNMAN once put it: *"One of the fundamental laws of physics is that electric charge is indestructible; it is never lost or created."* [19]

What we imagine here, however, is a field structure formation and energy transformation process which does not violate the principle of conservation of energy. This is exactly what the experiment demands from us, however without doubting the existence of a reality of whatever kind. This fact naturally changes our ideas about the

19   https://www.feynmanlectures.caltech.edu/II_13.html

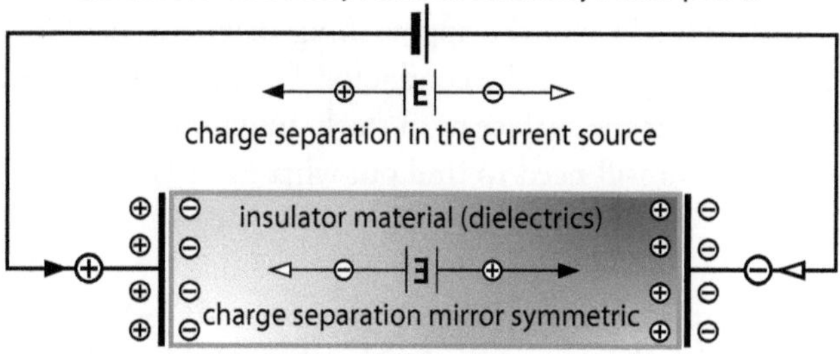

Long-term storage for electricity
(glass block, Leyden jar, ceramics, condenser, supercapacitor)

The applied voltage represents an electric field **E**, which generates a spatially inverted, mirror-symmetric antifield Ǝ in the glass block. Since the double-slit experiment clearly refutes particle ideas of all kinds, we can only explain this effect by holistic cell division and branching processes of an electric field that creates opposite charges **e⁻**‖**e⁺** (in mirror-symmetric notation: e ‖ ɘ). This is a scale-free, holistic process: the entire body is dielectrically polarized (local and global)

Figure 22 Dielectric polarization through branching processes

structure of atoms or integral molecules and leads to dynamic field structure models, very much in the spirit of FARADAY & MAXWELL. The dimagnetic polarization of magnetically non-conducting materials can be interpreted in the same way. We should therefore also be able to store magnetic energy in diamagnetic materials like water, wood or helium. Isn't that the way matter stores sunlight and heat? That the structure formation of molecules can indeed be explained by magnetic fields that can divide holistically will be discussed in detail in the second volume.

But back to our induction problem: Does the induced electric field branching takes place inside or outside the conductor? Since copper and silver are normally not magnetic, diamagnetism must also play a role in our question. Due to the counter-reaction of the diamagnetic material, the approaching magnetic field can most likely hardly penetrate the metal (since it is filled by the magnetic anti-field, which shields), so the surface of the conductor must be a surface on which a mirror symmetry between the outer and the inner magnetic field is formed.[20] This is reminiscent of the partial reflection of light.

20   Modern quantum theory states that the magnetic field penetrates the material and generates myriads of magnetic dipoles which represent structural changes in the atoms (our integral molecules) which collectively act as one whole (as one field). This could well be interpreted holistically.

So it is quite unlikely that the approaching magnetic field induces an electric current pulse *in* the conductor. Presumably, this process rather takes place at the surface or in the boundary layer. But before we can be sure, we still need to find out what ideas later researchers developed.

The similar formation of structures in magnetic, dielectric and conductive materials led FARADAY to the idea that electric and magnetic forces might be polarization processes, which could be interpreted as local structure changes (but a scale that defines locality is missing). He designed a model in which a field line consists of a series of electric or magnetic polarizable molecules or field states. When excited, they generate opposite electric or magnetic dipoles. Merging electric dipoles induce magnetic dipoles and vice versa, so that *polarization states*, not material units, are transferred from place to place. When these excited states propagate along a line, they can be interpreted as lines of force and field, but also as light rays and light waves. The assumption of an steady continuous motion automatically implies a time parameter, since one polarization state causes the next (which is opposite). In this way FARADAY established a connection between electromagnetism and FRESNEL's wave theory of light.

FARADAY's next discovery provided the crucial clue to the magnetic nature of light: In 1845, he found that a strong magnet rotates the plane of polarization of a light ray sent through a glass body. So the light ray had to be magnetic! This was another wonder of nature: Light can interact with magnetic fields, at least *polarized light*. This is the *magneto-optical* or FARADAY effect. What was not recognized then (and today) is the fact that this was an already branched and *therefore* polarized partial ray of light. In other words: The polarization of the light results from the branching of a light beam which thereby acquires magnetic dipole properties – the two partial rays are now magnetically opposite polarized. We will come back to this several times, because this branching process is the key to the true nature of light and matter (and to a true understanding of quantum physics).

Summary:

Faraday began to combine all these findings to explain the relationship between magnetism, electricity and light with the help of the field concept. For Faraday the field was a state of space which exists as real as ordinary matter and gases. In his model, there are no more instantaneous remote actions over arbitrarily large distances, but the field itself transmits effects and forces in the form of polarization processes, which propagate chain-like either spherically or radially. Basically, the entire field state describes a physical polarization state that can connect individual matter objects over arbitrary distances. The spatial subdivision of this polarization state can be arbitrarily coarse (at least two) or fine – that is only a question of the scale. These are the wavelengths. The smallest wave unit (in the energetic sense) consists of two opposite polarization states, which basically means that the two objects are connected by a dipole field structure that can be multiplied any number of times – which could also represent a stationary field with an arbitrarily dense structure.

Moreover, according to Faraday's conception, we should imagine solid matter no longer as substance-like solids, but as extreme local compactions of fields – and thus of polarized field structures. For Faraday, the atomic doctrine was *"at best an assumption whose truth we cannot confirm whatever we think or say about its probability"*.

Apparently, the limits of the atom hypothesis were still clear and doubts about the indivisibility assumption not yet extinct. He also suggested that gravity could be understood as a field that deforms space, just like magnetic and electric fields, and assumed a hidden connection between electric fields and gravity. If this were true, the acceleration of large masses should cause electric currents in conductors and dielectric polarizations in insulators. However, Faraday's first experiments did not succeed in proving such effects. Whether they exist has not been clarified until today.

Faraday's discovery of magneto-electric induction was one of the most important discoveries of physics. It provided the foundation for a growing understanding of electric and magnetic interactions and led to a new technology that should change our life forever. The induction principle led to the invention of electric motors, dynamos and transformers and caused thirty years later thousands of electro companies to spring up like mushrooms. FARADAY's discovery of the magnetic properties of light had the same significance, even though it had not led to technical applications yet (this is a case for emerging quantum technologies). By this finding it became clear that the true background of FRESNEL's transverse wave model was a magnetic dipolarity, as MALUS had already suspected. That polarization had to do with the bifurcation of light beams was obvious, but the composite hypothesis of the wave theory prevented the insight that the branching process had to be a physical structure-forming process.

So, as it looks, FARADAY's findings do not argue against interpreting electric and magnetic polarizations as holistic division and branching processes of matter and fields either. But before we can be reasonably sure, we still have to study what MAXWELL, WEBER and HERTZ have to say about the nature of matter and light. By the way, FARADAY did not yet need an exaggerated use of mathematics. His physical and chemical insights were based solely on pictorial ideas that he developed from practical experiments.

This was about to change.

Maxwell's field theory: Gaseous matter in motion

Between 1855 and 1864 JAMES C. MAXWELL (1831-1879) attempted to translate FARADAY's discoveries, his field concept and all known facts about electricity, magnetism and light into an all-encompassing mathematical theory. He had already produced the world's first color photograph in 1861, following the hypothesis of YOUNG (the inventor of the double-slit experiment) that human color vision is based on only three receptors in the human eye that react to red, green, and blue light. YOUNG had already discovered this in 1794, at the age of twenty-one. We still use this RGB principle in color printing and computer screen technology today.

To unify electromagnetism and light, MAXWELL had to clarify how the findings of COULOMB, OERSTEDT, AMPERE, GAUSS, FARADAY, WEBER and many other physicists could be combined with the field concept and the transverse wave model of light of YOUNG and FRESNEL. However, it was not yet clear exactly what electric charges are and what an electric current really is – two electric fluids, electrically charged molecules, electric particles, two counter-currents or something else – and whether the direct electrodynamic force or the polarized magnetic field around the conductor is the better model. Nobody knew how polarized light comes about and how the magnetic properties of polarized light waves can be explained. Decisive was the conceptual interpretation of the key experiments of YOUNG, OERSTED, FRESNEL, AMPERE and FARADAY. However, the crucial question was whether the interaction between two matter structures can be understood as an instantaneous action over arbitrarily large distances, as in the theories of NEWTON, COULOMB and AMPERE, or whether physical actions were mediated by local excitations of a gaseous fluid of extremely low density, as had been suspected since HUYGENS, YOUNG and FRESNEL.

Initially, between 1855[21] and 1862[22], MAXWELL attempted to illustrate the interaction of electricity and magnetism with the help of mechanical models inspired by gear and power transmission principles. On the basis of such analogies he derived several mathematical equations that intended to model the experimentally facts. His orig-

21 James Clerk Maxwell, "On Faraday's Lines of Force" (1855)
22 James Clerk Maxwell, "On Physical Lines of Force" (1861/62)

inal idea was to derive regularities from the observable mechanical motions caused by electric and magnetic forces on bodies. At the same time, he tried to interpret FARADAY's field lines as tensile and compressive stresses and elastic strains of a deforming gaseous medium, which should produce the electric and magnetic forces. He compared FARADAY's field lines with streamlines of fluids and vortices and with mechanical stress lines in common materials, as they were already known from statics and dynamics in engineering and material sciences. He was also inspired by experiments with translucent gelatin blocks, which allowed him to visualize internal mechanical stresses and torsional motions by shining light through them. With this approach he followed the ideas of WILLIAM THOMPSON (later LORD KELVIN), who had tried to explain FARADAY's magneto-optical effect, the rotation of the plane of polarized light by magnetic fields, as an elastic torsion of a magnetizable aether gas.

MAXWELL was also familiar with the work of the German physicists WILHELM WEBER (1804-1891) and RUDOLF KOHLRAUSCH (1809-1858). It was their groundbreaking work that inspired MAXWELL to his famous insight that light must be electromagnetic in nature. A few years later he studied the work of HELMHOLTZ, who had used one and the same equation as a universal flow model for hydrodynamics (water flow), heat propagation (thermal flow) and electric potentials (electric flux) in 1858. However, this was a rather unfortunate analogy that sometimes led physicists to confuse qualitative properties with quantitative aspects, although positive and negative charges must be primarily understood as opposite physical *qualities*. Since then, electric current has been imagined to flow from a higher energetic potential (plus) to a lower one (minus), just as water flows down a hill or heat from warm to cold. HELMHOLTZ also designed the very first mathematical model of the vortex motion of fluids, for which he introduced streamlines inspired by FARADAY's field lines. However, at that time it was still completely unknown that streamlines must be granted the property of divisibility, since all flow systems are subject to non-linear dynamics, which in principle cannot be calculated. In other words, streamlines can branch or bifurcate. However, this was only recognized in the context of nonlinear complex dynamics at the end of the 1970s.

HELMHOLTZ' vortex theory, AMPERE's interpretation of the polarity of the magnetic field around current-carrying conductors as circular motion, and his own interpretation of the electric current as unidirectional motion of negative charges (in contrast to AMPERE) apparently led MAXWELL to believe that the magnetic field around an current-carrying conductor is actually circulating. However, this remains pure fiction. These pictures are only mechanical analogies which MAXWELL had designed to be able to model the polarization movements mathematically at all. From these mechanical analogies he derived his field equations which claim to correctly capture the interactions of electric and magnetic fields as dynamical deformations of a hypothetical aether. This is the next switch of cognition which shows how much physics depends on fundamental conceptions.

First, MAXWELL sketched out a rough idea of what he actually wanted to depict. He had to consider that gases and solids are electrically and magnetically polarizable, that *somehow* moving negative electric charges make an electric current which in turn induces a magnetic field with a particular rotational sense (magnetic polarity), and that polarized light must have magnetic properties. Moreover, it seemed certain that light propagates in the form of transverse waves. MAXWELL now suggested that transverse waves are electric and magnetic polarizations traveling through a gas called the aether. Therewith he presupposed the validity of the continuity assumption, which is the precondition for the application of continuous or linear differential equations. Thus, the aether was supposed to be an isotropic, continuous gas, uniformly constituted. The idea of self-structuring or partitioning of a gas by structure-forming holistic division processes was not yet in the air, as we have already seen in the discussion of AVOGADRO's molecular division hypothesis. Nevertheless, and this is astonishing, MAXWELL already brought into play the concept of a cell-like structure of the gas (in a rather obscure way). However, he did not think of ordinary gases, but of the aether gas, which apparently should have a special position among the elements.

Originally, the magnetizable aether[23] was to have a structure consisting of 'cells' or 'vortices' traversed by a capillary network of electric fluid. This fluid was to consist of negatively charged particles or volume elements whose displacement represented the electric current. The motion of these electric charges should cause these cells to rotate by friction, creating magnetized cells or vortices.[24] In this way, the kinetic energy of the electric charges should be stored in the aether gas in form of magnetic energy. When the process is reversed, the opposite rotation of the cells shifts the electric particles back to their initial position (this picture serves to model opposite electric polarizations and current directions). So the real goal was to design a mechanical model (with moving bodies) that can represent FARADAY induction and the findings of OERSTED and AMPERE as reversible interactions between electric and magnetic polarization states. In order to explain the complete reversibility of the magnetic storage of electric energy in the aether gas, which was an indispensable condition for a continuous propagation of light waves, MAXWELL assumed an ideal elasticity of the gas cells (by the way, this must be valid also if we assume structural changes of the gas molecules by holistic division processes). However, ordinary gases and insulating materials – both of which are dielectrics, i.e. electrically non-conductive subsstances – do not behave ideally with regard to the complete reversibility of energy storage. This was already known from the Leyden jar. In dielectrics like glass, residual charges remain; the discharge process itself oscillates over a long period between opposite electrical polarization states.

23   In this book I have adopted the original spelling of Maxwell (1864), as it was and is common in German. This allows a clear distinction from the chemical substance called ether.

24   This picture was apparently intented to model the magnetic polarization as circular motion as in Fig. 15 and 16 (p.111 ff). Why he then drew hexagons as rotating cells will probably remain his secret forever. However, the real reason for the introduction of the "idle wheels" modeling electricity, which he interpreted on the one hand as electric particles and on the other hand as "incompressible electric fluid", was the fact that according to Ampere the transverse magnetic field lines or "magnetic vortices" must all rotate in the same direction if the currents run parallel in the same direction. However, this is not possible if the vortices are in contact and through this very contact the motion is to be transmitted continuously and without loss. In mechanics, a rotating friction wheel or gear wheel would not turn the contacting neighboring wheel in the same direction, but in the opposite direction. Therefore, Maxwell had to insert "idle or friction wheels" into this picture, which he then interpreted as "electrical components." In other words, the mechanical metaphors provided only a cumbersome, barely visualizeable, and rather crude model (for Maxwells original drawings see Wikipedia).

The following quote from the 1862 paper "On Physical Lines of Force" shows how MAXWELL explained his mechanical analogies:

"According to our hypothesis, the magnetic medium is divided into cells, separated by partitions formed of a stratum of particles which play the part of electricity. When the electric particles are urged in any direction, they will, by their tangential action on the elastic substance of the cells, distort each cell, and call into play an equal and opposite force arising from the elasticity of the cells. When the force is removed, the cells will recover their form, and the electricity will return to its former position".
(On Physical Lines of Force, 1861, Part III, 15)

In this paper he already presented the basic idea for his final theory, which led to the unification of the electromagnetic theory with the optical theory of light. He came across it when, while studying WEBER's equations, he noticed a suspicious connection between two numbers that seemed to have nothing to do with each other: The first number was WEBER's constant ratio between electrostatic forces and electrodynamic (magnetic) forces in current-carrying wires, the second number was the velocity of light, which had already been determined in optical experiments. Since WEBER's ratio was determined experimentally from the electrostatic forces and the current intensity of the discharge through a conductor, it is independent of the underlying physical ideas of what an electric current really is.

Since WEBER assumed that the electric current consists of opposite electric half-charges moving past each other in opposite directions, he interpreted his constant c as a *relative speed* between these half-charges, or between a moving half-charge and the conductor. MAXWELL now realized that this speed would correspond to the velocity of light if he modified the underlying physical ideas a little bit. For this he would only have to replace the bidirectional current concept of AMPERE and WEBER by a *unidirectional motion of negative full charges*, which would mean that the electric current consists of only one kind of electric charges. He continued to pursue this idea, although he was aware that the question of what an electric current actually is had not yet been settled at all:

"We have, in fact, now come to inquire into the physical connexion of these vortices with electric currents, while we are still in doubt as to the nature of electricity, whether it is one substance, two substances, or not a substance at all, or in what way it differs from matter, and how it is connected with it." (On Physical Lines of Force, 1861, Part II, 285)

This quote shows that MAXWELL actually interpreted the polarity of the magnetic field around current-carrying wires as a rotating magnetic field. Sometime between 1862 and 1864, however, he realized that it was difficult, if not impossible in principle, to model electrical and magnetic interactions and the associated structural changes of the gas due to polarization processes with mechanical motions. With no new, non-mechanical ideas in sight (no one had followed up on AVOGADRO's non-atomic approach to molecular gas theory), he finally abandoned his goal of designing convincing pictorial-spatial ideas of these structural changes.

Therefore, in his "Dynamical Theory of the Electromagnetic Field" of 1864, he restricted himself mainly to the design of mathematical equations which reflected relationships between measurable quantities, e.g. between electric and magnetic charge units, electric and magnetic forces, or between voltage, current strength and electrical resistance. The determination of the units was largely based on the experimental works of GAUSS, AMPERE, WEBER and KOHLRAUSCH, but MAXWELL determined these values additionally by own measurements. Guided by the transverse wave theory of FRESNEL and the induction principle of FARADAY, the velocity of light in optics, and WEBER's equations, MAXWELL developed the idea that light and heat radiation can be understood as electromagnetic waves consisting of polarization excitations, which can be understood as dynamic deformations of an extremely rerefied gas, the aether. This primordial gasous matter should fill the entire universe and permeate all material bodies. The waves are traveling, cyclically emerging and decaying electric and magnetic polarization states whose propagation speed is identical with the speed of light in optics. This speed is also identical with Weber's constant when multiplied by two. In this way MAXWELL's field theory of 1864 describes gaseous *"matter in motion, by which the observed electromagnetic phenomena are pro-*

duced". Since the deformations of the aether gas generate forces and mechanical movements (pressures, tensions, torsions and change of location), he understood his field descriptions as a *dynamic* theory of matter.

MAXWELL always held that the mathematical field model is only a means to describe the topological deformations of a real existing matter. However, he had to forego a real physical understanding of structure changes, since he could only paraphrase the phenomenon of polarization with mechanical metaphors of motion. Despite this difficulty, he succeeds in presenting the relationships between electric and magnetic effects in an understandable way. If one studies his work of 1864, one is surprised how clearly he explains the state of physics and his concept. MAXWELL's theory of electricity, magnetism and light therefore appears coherent and understandable – at least at first glance. Remarkable is that he practically uses only one physical principle to describe the electric and magnetic properties of matter and light, namely *polarization*. So it seems that MAXWELL was the first to fully realize the importance of polarization processes for our understanding of the structure of matter and light.

The problem is only that polarization, which creates a dipolar physics in the first place, must be a *non-mechanical motion* whose true nature he was unable to fathom. Precisely this leads to considerable difficulties at second glance – making a genuine understanding of the polarization process almost impossible. In my understanding, it is actually about dipolarization, i.e. the formation of dipolar, exactly opposite physical properties in matter and light, or enantiomorphic fields in general. This is what MAXWELL's theory cannot map. One reason is the particle concept, the other one is the transverse wave theory. Both exclude holistic division and field branching processes from the outset and therefore cannot explain why even single rays or waves of light (and matter) can divide and still remain a whole. When this difficulty became clear forty, fifty, and sixty years later, via convoluted detours, one was forced to develop quantum theory – again without understanding the underlying problem: The existence of polarizing branching processes.

Maxwell's mechanical models of polarization

MAXWELL's equations are based solely on analogies to mechanical motions, which were intended to model polarization phenomena. He represented electric polarization as linear displacement of negative charges and magnetic polarization as rotation of aether cells, molecules, or magnetic vortices. The latter are modeled as rotation in one of two possible opposite directions, the electric polarization as unidirectional motion of exclusively negative charges in one of two opposite directions. A polarization process in the original sense of the word, i.e. a *simultaneous* dipole formation, was thus not considered by MAXWELL, although this would have fitted well to the bidirectional current model of AMPERE and WEBER. To explain how the negative electric charges are set in motion and an electric current in the conductor is generated, he combined AMPERE's conception of the *electromotive force* with FARADAY's magnetic induction:

"This, then, is a force acting on a body caused by its motion through the electromagnetic field, or by changes occurring in that field itself; and the effect of the force is either to produce a current and heat the body, or to decompose the body, or, when it can do neither, to put the body in a state of electric polarization – a state of constraint in which opposite extremities are oppositely electrified, and from which the body tends to relieve itself as soon the disturbing force is removed."

When reading this description, one realizes again that the "electromotive force" can also be interpreted as a process which separates positive and negative charges, i.e. creates opposite charges in the first place. If we interpret the electromotive force as a process, we recognize in this description a movement which occurs simultaneously in two directions. We interpret these as a branching motion, which generates a dielectric polarization (which appear as opposite electric charges). So, in this new picture we have to do with a field-branching process, a holistic division, by which a new field structure with electric dipole properties and ontological-mirror-symmetrical properties arises. This enantiomorphic field is what we call an "electric field". In this sentence we can see that our model of the branching of a – then electric – field could explain the facts just as well. The two opposite electrified "extremities" of the body are caused by the two opposite polarized field parts. However, in MAXWELL's model only negative electric charges move to one end of the body, resulting in a void at the other end, which appears as a positive pole in a kind of *contrast perception*. However, in our new model, the *process* of branching generates an enantiomorphic electric field, which is identical to the electromotive force. It is also identical to concepts such as the generation and separation of opposite charges, dielectric polarization, the formation of electric dipoles, the electric field, and a bidirectional current pulse. The reversal of the electric polarization process is a fusion of this field structure element to the original symmetry state. This is a clear difference to MAXWELL's conceptions.

It is noticeable that MAXWELL uses many terms ambiguously and yet always describes the same principle, such as *electricity* both for charge units and a fluid medium, a charge displacement in the mechanical sense, a displacement current in the field theory sense, an electric polarization for conductors, and a dielectric polarization for insulator materials, and so on. Ultimately, however, all these terms describe either electric or magnetic polarization processes of matter, gases and the hypothetical aether gas. The use of different terms for one and the same physical principle is obviously due to the different manifestations of that principle. This has confused many physicists and probably also prevented them from grasping polarization as an elementary structure forming process. But the main reason was apparently that he could not yet clearly recognize the non-mechanical nature of polarization. Nevertheless, he was convinced that polarization must be related to the states of a real existing gaseous matter in the universe:

"As difficult as it may be for us to gain a conclusive idea of the nature of aether, there can be no doubt that the interplanetary and interstellar spaces are filled with a material substance or body".

However, it remained unclear what type of substance the aether gas should be. He only knew that mechanical ideas and a wave motion alone were not sufficient to grasp the true nature of the deformations of the gaseous matter *"by which the electromagnetic phenomena are produced"*. For this reason, from 1864 on, he warned his readers that the underlying, apparently pictorial mechanical models for the motions with which he described the polarization phenomena *were only analogies without any claim to correctness*. They had been only auxiliary models which finally led to his field equations:

"I have on a former occasion attempted to describe a particular kind of motion and a particular kind of strain, so arranged as to account for the phenomena. In the present paper I avoid any hypothesis of this kind; and in using such words as electric momentum and electric elasticity in reference to the known phenomena of the induction of currents and the polarization of dielectrics, I wish merely to direct the mind of the reader to mechanical phenomena which will assist him in understanding the electrical ones. All such phrases in the present paper are to be considered as illustrative, not as explanatory." [25]

This statement did not only mean the rough mechanical idea from 1855, in which rotating magnetic vortices or 'cells' and electric particles should be imagined as force mediators. It refers mainly to the assumed *unidirectional motion* of negative charges and the *apparent rotation* of the magnetic vortex around current-carrying wires. The problem is only that MAXWELL tended to conclude from 1862 on that an electric current *actually* consists only of negative electric charges and flows only in one direction, although, as he explicitly notes in every paper from 1855 to 1864, this question had not yet been settled at all.

Summary:
MAXWELL rejected direct and immediate forces in favor of the FARADAY field concept, replacing them with local polarization processes and energy transfers that would cause deformations of an extremely dilute, homogeneous, and gaseous medium. However, the exact meaning of terms like *location*, *local* and *near-action* has never been clearly defined (as we will see, this is a question of the scale applied).

25 James C. Maxwell: „A Dynamical Theory of the Electromagnetic Field", 1864, S.461
Philosophical Transactions of the Royal Society of London, Vol. 155 (1865)

For MAXWELL, the main reason was that light and heat energy have been shown to take time to propagate. This is explained by structural changes in the hypothetical aether gas, identified as electric and magnetic polarization motions and modeled with continuous time cycles. This gas should be magnetically and electrically polarizable like ordinary gases and matter, so that the same rules apply to matter on Earth and in the universe. In this sense MAXWELL designed the first unified theory of matter and light. The insight that matter and light must have the same nature was neither obvious nor easy to understand – in other words, simply grandiose.

We have heard this so often that we are hardly surprised. But if one really thinks about what such different phenomena as positive and negative electricity, electric currents, magnetism, dipoles, polarization, structural changes of gases and light are supposed to have to do with each other (and that they are all somehow the same thing), even seasoned physicists get into trouble when they have to explain that. Since the aether concept was abandoned over a hundred years ago (at EINSTEIN's suggestion, 1905), many physicists today no longer know that electromagnetic field theory was originally a model in which light was just a manifestation of an very elastic gas in structural motion – a gas with extremely low density, almost a vacuum (but not a perfect one), which nevertheless was supposed to behave like a coherent gas volume. Since MAXWELL could not find a convincing non-mechanical idea for the polarization motion, he finally gave up any attempt to find a conclusive physical model for these movements and considered electric and magnetic interactions exclusively in terms of mathematical relationships between measurable quantities. These relationships are based on the connection of the field description with the concept of motion of mechanics, which always presupposes a directional motion of material objects or energetic states. Some physicists and electrical engineers today also do not know that the theory originally consisted of twenty differential equations. The reduction to four equations, as we know them today, took place only decades later by HERTZ and HEAVISIDE. And it may well be that an important idea has been lost in this transformation...

What has been lost is MAXWELL's concept of magnetic polarizations in the aether gas, or of a 'free' electromagnetic field in a vacuum, if the aether hypothesis is abandoned. The idea, that magnetic polarizations can occur not only in ordinary gases on Earth, but also in the universe, was already banished from the electromagnetic theory by HERTZ and HEAVISIDE. And with it was also eliminated the possibility of thinking that free electromagnetic fields can produce dipolar magnetic properties, which we call *spin* today. Accordingly, it proved difficult later (and still today) to understand the sudden appearance of 'spin' as a structural property of branched fields, even though the experiments show a division of light rays and molecules that simultaneously produces exactly opposite, i.e. dipolar, magnetic properties in the two partial waves. This again raised the question of what magnetic polarization or 'spin' actually is: Is it a property that light waves and 'elementary' particles possess from the outset, or is it a new physical property that only arises through interactions, also called measurements? The problem has not been solved until today, which has directly to do with the prevailing view about the unidirectional motion of light and the existence of negative electric particles.

In our interpretation, however, the magnetic dipolarization of a gas or of an electromagnetic wave would mean that either a molecular cell division of the gas or a field branching of the light must take place, where the two connected cells or branches assume opposite magnetic properties (they show exactly opposite vectors of the magnetic field strength, which is the 'spin'). If this is the case, magnetic, electric and electromagnetic fields can divide holistically and form enantiomorphic structures that are scale-invariant, i.e. can exist in any size (so it is a physical quality, not a quantity). This is consistent with our interpretation of AVOGADRO's hypothesis as molecular cell division, which also requires scale independence. Thus, in this model, the division of an integral molecule creates the exactly opposite magnetic properties of the halves, while the molecular structure may have an indefinite size. And the same applies to branched light. The question is then, how scales and certain size relations in nature come about at all. EINSTEIN's special theory of relativity will help us to clarify this question.

MAXWELL's work shows that the problem of finding the right physical conception for electric and magnetic polarization processes was originally directly related to the question of the true nature of gases. At this point, our new idea can shed new light on the old question of what *gaseous matter in structural motion* really means. There is no reason to be afraid of such a question, even if EINSTEIN advised us that it is best not to mention the irritating A-word anymore, because this question refers first to quite normal gases. That gases are electrically and magnetically polarizable is out of question today. That the same gases also exist in the universe is nowadays also clear. Likewise the fact that their density is so low that we can speak of a vacuum, as MAXWELL correctly assumed. Do you even know how many molecules a high-vacuum of modern physics contains? MAXWELL and EINSTEIN could not know that yet: *10 billion molecules* per cm^3! And the interstellar gaseous medium still contains about *1 million molecules* per cm^3. Of course, the density of a gas on Earth is enormously greater: For example, the density of air at sea level is estimated to be 10^{19} molecules.[26] So you see, emptiness is a pretty relative concept. Moreover, we know that a low density facilitates the magnetic polarization of gases. So whether a gas in the universe exists that is magnetically polarizable and structurable as a whole – and thus can bridge arbitrarily large volumes and distances – is a question of the proper gas theory and its kinetics. That the kinetic gas theory of mechanics cannot be correct is not only shown by the double-slit experiment with single molecules and atoms. By the way, one of the most important approaches to a new, holistic understanding of the molecular theory of gases and its kinetics comes from EINSTEIN – hidden in the theory of the monatomic gas, which was *helium*. We will return to this topic once we know more about the molecular orbital theory of quantum physics, which gets along without atoms.

Until this question is settled, our new concept can be applied directly to the structure formation of electromagnetic fields, because the main problem of field theory is actually the same: To understand polarization (aka "spin") with a new physical principle that applies equally to gaseous matter and light. As the infamous wave-particle paradox proves, the not yet understood nature of polarization still prevents us from understanding the true constitution of reality until today.

26   https://en.wikipedia.org/wiki/Interstellar_medium

With the phenomenon of polarization, in the first half of the 19th century physicists came across a new form of motion, still unknown to physics, which could no longer be understood as motion in the sense of mechanics. It revealed itself in the structural properties of light and matter and had to be responsible for their electrical and magnetic properties. If our hypothesis is correct, it is about motion in the sense of structure change, about holistic division processes, which manifest themselves as magnetically polarized branching of light and molecular cell divisions.

Pursuing this idea further, it is easy to see that the bidirectional electric current could be something quite similar, namely a branching process that makes a holistic, intrinsically opposite electric field, i.e., a dielectric polarized field structure. Thinking further in these direction, one can also recognize that the holistic divisibility of light corresponds to the holistic divisibility of gaseous matter. The only obstacle preventing this kind of physical thinking is the unidirectional definition of both the electric current and the speed of light. So here, in the foundations of electromagnetic theory, another fork in the path of cognition opens up, which is crucial for understanding the like physical nature of light and matter:

Weber: The bidirectional electric current and the relative speed of light

Now we come to the most interesting part: We investigate the suspicion that something might be wrong with our understanding of the speed of light. So let's take a closer look at how MAXWELL discovered that the speed of light known from optics also appears in WEBER's electrodynamics, that is, in the theory of electric current.

Today it is hardly known that MAXWELL's famous discovery that light is electromagnetic in nature and propagates at a constant speed is based directly on the work of WEBER and KOHLRAUSCH. MAXWELL was well acquainted with the research of WILHELM WEBER, who in 1855 had experimentally derived the first exact definition of absolute units of measurement for electrostatic and electrodynamic (magnetic) forces, and thus, without noting it at first, the speed of light from purely electrodynamic principles. Nowadays, few physicists and electrical engineers know the name WEBER, and even fewer know that he had derived this speed on the basis of AMPERE's bidirectional current concept – although MAXWELL had explicitly referred to WEBER's research in his papers of 1855, 1862 and 1864. We even do not know if EINSTEIN knew WEBER's original work and his definition of the constant c as a *relative* speed. Even the famous RICHARD FEYNMAN, who influenced generations of physicists with his brilliant lectures, does not seem to have known WEBER's derivation of the constant c. Given the seemingly insurmountable problems of understanding that have arisen in the physical interpretation of MAXWELL's theory, special relativity, and quantum physics, this lack of historical knowledge is quite astonishing. Actually, it is not difficult to see that all these cognitive problems are related to the unsolved issues of MAXWELL's theory: What exactly is electric and magnetic polarization? Is it motion in the sense of mechanics, i.e., bodies, particles, or states changing their location, or motion in the sense of structural change? What is electric current? Does it consist of positive and negative electric fluida or only of negative charges? Is it a unidirectional motion or a bidirectional one? Is it a field branching process? What may seem surprising at first glance is that any answer to the question of the nature of electric current also affects our understanding of the nature of light – and thus our *definition* of the speed of light.

Although MAXWELL admired and greatly appreciated the research of WEBER and KOHLRAUSCH, he did not share their interpretation of electric and electrodynamic (magnetic) actions, since they were supposed to act instantaneously over arbitrarily large distances, as in NEWTON's theory of gravity and AMPERE's electrodynamics. Their model required direct and instantaneously acting forces (actions at a distance) between electric charges and bidirectional electric current elements as a function of their electrical polarization, relative angle, relative speed and acceleration. This was basically in accordance with COULOMB's law, except for the acceleration term, which does not appear in the electrostatic law. In the concept of WEBER, an electric current starts with a charge separation in the battery, a process that FARADAY had thoroughly explored and explained with the migration of oppositely charged molecules (ions) to the respective oppositely charged poles inside the battery. In a closed circuit, charge equalization then takes place as the positive charges move via the outer conductor from the positive pole to the negative pole and the negative charges from the negative pole to the positive pole. In this model, therefore, the electric current consists of two opposite electric *half-charges* moving past each other in opposite directions. They are considered as half-charges, since their electric symmetry must result in zero due to charge separation. So this is a qualitative statement, not a quantitative one (Fig. 24 and 25). This was a view that seemed well founded experimentally and was shared by many other scientists. Shortly after MAXWELL had begun to think about the deformations of the aether gas by electric and magnetic polarization processes, he discovered in WEBER's work of 1854 the decisive clue to the commonality of light and electric current.

WEBER and KOHLRAUSCH had already solved a whole series of problems. The first problem was to determine experimentally the absolute units of the electric and electrodynamic force, the second was to relate these quantities to each other, and the third was to define a unit of electric current as the quantity of electric half-charges moving in a conductor producing a certain electrodynamic (aka magnetic) force. The starting point was that the intensity of electric current can be determined in various ways and expressed in different units. For example, in electrolytic measurement, where water splits into hydrogen and oxygen by applying a voltage, the intensity of the electric current is expressed as the dissociation rate of one milligram of water per second as unit. But if the action of an electric current is measured by the deflection of a magnetic galvanometer needle, the intensity is expressed in absolute magnetic force units that produce a certain torque. For this reason, there are different absolute units for electrostatic, magnetic and electrodynamic determined current intensities.

KOHLRAUSCH had already investigated electrostatic actions using a Leyden jar as electrical storage device. He had precisely determined the quantity of electricity that can be stored in the glass material, the magnitude of the repulsion force of a unit electrostatic charge (which was divided into two equal amounts for this purpose), the magnitude of the attraction force of opposite electric charges, and

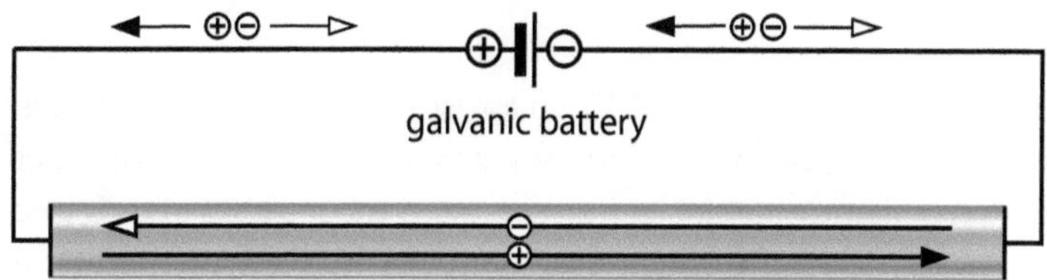

Ampere's bidirectional current (e⁺ + e⁻)
The lower segment of the wire is shown as an enlarged cylinder (gray)

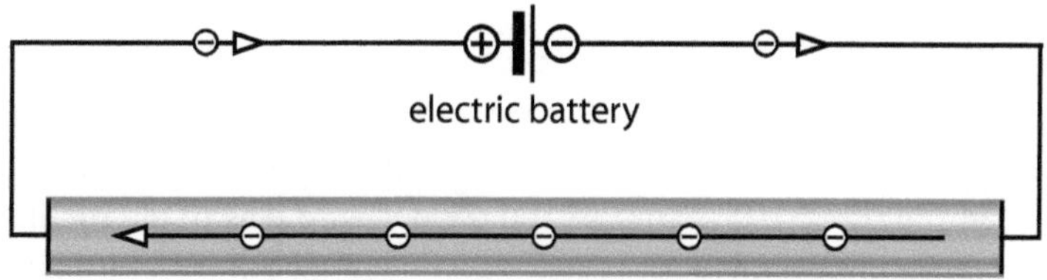

Maxwell's unidirectional current (e⁻)
consisting of only negative charges moving from minus to plus

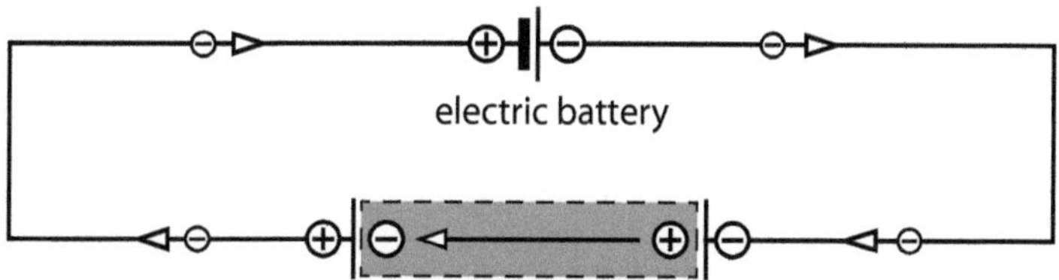

Maxwell's displacement current
open circuit with gap, filled with arbitrary dielectrics
(air, aether, capacitor)

Figure 23 Electric current models of Ampere and Maxwell

the magnitude of the electrodynamic (magnetic) force with which a unit electrostatic charge acts on the compass needle of a precision galvanometer during discharge. These measurements exactly determined the electrostatic force unit according to COULOMB and the magnetic force unit according to GAUSS. As early as 1846, WEBER had already derived a universal law of electrical action and determined the electrodynamic force unit according to AMPERE, which gives a measure of the strength of a defined electric current flowing through a wire of a defined length per unit time during discharge.

At Easter 1854, Weber and Kohlrausch together repeated these experiments in Göttingen, Germany, as accurately as possible. Their goal was to determine the relation between the number of unit electric charges flowing through the conductor and the magnitude of the electrodynamic force induced by them. Recall that the electrodynamic force is nothing more than the magnetic force induced by electric currents. So both terms mean the same thing – the magnetic force with which an electric current acts on a magnetic needle. The only difference is that they are expressed in different absolute units that can be converted into each other if their ratio is exactly known. The determination of the electrodynamic force (or strength of the induced magnetic field) around the copper wire was made by discharging a Leyden jar with very high voltage (about 30.000 volts) through a very long isolated copper coil. The discharge pulse of the exactly determined electric charge induces a certain electrodynamic (electromagnetic) force, which acts on a special precision galvanometer. The result of this experiment was a precise determination of the numerical ratio between the electrodynamic unit force and the electrostatic unit charge. According to Weber, this numerical ratio is a *constant* that expresses how many unit electric charges per second must flow through the wire to produce an electric current with the electrodynamic defined current intensity of 1 Weber (Wb). The ratio to the modern unit of current intensity 'Ampere', which is expressed in electromagnetic measure, is 10 Wb = 1 A.

Of course, this is only a simplified description of rather complicated, extremely sophisticated experiments that are of groundbreaking importance for physics.[27]

The system of Weber and Kohlrausch for the determination of the current intensity through electric charge units was state of the art and brilliant, because it does not depend on the exact number of the electric charges in the electrostatic unit charge. The only condition was that exactly the same amount of electricity was used as standard unit for these combined measurements. Note: At that time

27   Detailed descriptions of that procedure are published by Weber and Maxwell himself. Overviews and discussions can be found in the works of A.Koch Torres Assis, K.H. Wiederkehr, and U. Stille, or in Ostwald's "Classics of the Exact Sciences", Vol. 5 (see bibliography).

electrons were still unknown – and not needed. This is only a discrete indication that the apparently dramatic failure of the particle concept and the atomic world view in the double-slit and interference experiments of quantum physics does not change the fundamental findings of electrodynamics. But really remarkable is that the constant ratio between the electrodynamic and electrostatic unit defines not only the current strength (i.e. the intensity of the polarized magnetic field around the conductor). Moreover, this ratio also represents a constant expressing a state of symmetry between electric and magnetic forces. It is a force-free symmetry state within a dynamic physical system. And as it soon turned out, this applies to both electric currents and light.

In WEBER's bidirectional model of the electric current, an electrostatic unit charge E consists of two opposite electric half-charges 1/2 e$^+$ and 1/2 e$^-$ moving in opposite directions. The constant numerical ratio represents a state of relative motion of two electric masses in which electrostatic forces are exactly balanced by oppositely acting electrodynamic (or magnetic) forces. For example, the electrostatic *repulsion* between two half-charges of the same type is exactly neutralized by a electrodynamic (magnetic) *attraction force*. However, in the case of two opposite half-charges, the electrostatic *attraction force* is exactly neutralized by an electrodynamic (magnetic) *repulsion force.*

So in (or at) the current-carrying conductor the electric attraction of the halfe-charges is balanced by a magnetic repulsion, and electric repulsion by magnetic attraction. To achieve such a force-free dynamical state, the half-charges must move at a certain speed, at least according to WEBER. He interpreted this constant, which he simply called c, as a speed, because the numerical ratio showed the 'dimensions' of length and time. This is characteristic for *absolute units*, which are free of any arbitrariness, since they are based exclusively on proportions between measured lengths, time intervals and accelerated masses. This is the so-called cgs measure system introduced by GAUSS and WEBER; the units of measurement are centimeter, gram and second.

Since the numerical ratio between the passing positive electric half-charges and the electrodynamic unit of force was 155,000 to 1, the speed of a half charge had to be 155,000 km/sec.[28] An interesting question now is: What you would recognize in this ratio if you were confronted with it for the first time? Is there possibly a completely different physical interpretation that has nothing to do with velocities?

MAXWELL, at any rate, noticed a suspicious coincidence of this ratio and the speed of light, which FIZEAU had determined experimentally with 315,000 km/sec in 1849. In 1851, FOUCAULT had determined the speed of light with an improved experimental setup and found a value of 298,000 km/sec. In this context, MAXWELL recognized that WEBER's c was about half the speed of light. This inspired him to double WEBER's constant c, which gives about 300,000 km/sec. Since an electric unit charge consists of two electric half charges $1/2\ e^+$ and $1/2\ e^-$, which move in opposite directions past each other, the speed of 300,000 km/sec denotes a *relative speed* between two opposite half-charges. However, this is not the interpretation of MAXWELL; it is WEBER's (Fig.24).

In contrast, MAXWELL interpreted this speed as the speed of a negative electric full charge moving in one direction only. This is exactly the point where the path of cognition branches out: He interpreted the speed of his negative charges as a *unidirectional motion* in the same way the propagation of light was understood, which is called *velocity* in physics. Apparently, he had recognized a deep commonality in the equivalence between the optically determined speed of light and the electro-dynamically defined speed of electric current: Both had to be similar in nature. To be able to represent this similarity, he replaced the bidirectional current concept of AMPERE and WEBER by a unidirectional electric current concept. Thus he adapted the theory of electricity to the theory of light. And there is the rub.

28 The exact value was 155,370 km/sec. According to Ampere, it is sufficient to define the polarization of the electric current and thus the current intensity by the direction of the motion of the positive charges. That this number must be multiplied by two, if the negative charges are also considered, was self-evident in this theory. However, Weber had initially overlooked the fact that the factor 2 reveals the identity of his constant c with the speed of light.

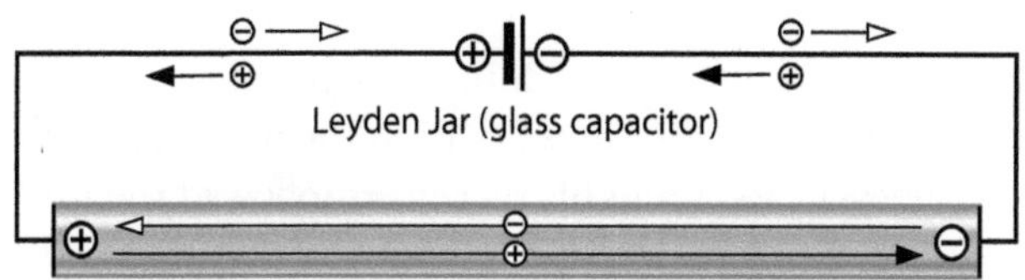

Weber's bidirectional current (½ e⁺ + ½ e⁻)
relative speed between a half-charge and the conductor: 150,000 km/sec
relative speed between opposite half-charges: 300,000 km/sec

Figure 24 Weber's bidirectional electric current and the relative speed of opposite half-charges

There is another solution which has not been considered yet: If there are doubts about the correct understanding of the nature of light, as suggested by the EPR paradox and quantum physics in general, one can also think about the need to adapt the theory of light to the bidirectional theory of electric current! This is possible as soon as polarization is interpreted as a fundamental physical principle of the nature of light, matter and electric current and identified with a field branching process.

Obviously, there was not the slightest doubt that light moves only in one direction. This should be true not only for single light rays, but also for a spherical wave front, although it expands in all directions simultaneously. Note that in a spherical wavefront one can always find two light rays or waves moving in exactly opposite directions. However, in MAXWELL's wave theory, which had assimilated FRES-NEL's transverse wave model and the composite hypothesis, there is no physical connection between them. They are always treated as separate waves and individual light rays moving away from the origin in opposite directions. But if these two assumptions are wrong, the EPR paradox is as good as solved.

If the speed of light is equal to the speed of electric charges (if the resistance of the wire is zero), then light and electric current could in principle have a similar physical nature. But exactly this insight opens up two possible ways of thinking: One can adapt the model of bidirectional electric current to the assumption that the propagation of light is unidirectional, as MAXWELL did (this assumption is shared by all physicists today), or one can try to capture the nature of light with a bidirectional branching motion and thus reformulate Weber's concept of electric current within the framework of a field theory. While Maxwell and his successors followed the first path, we will explore the second – a path, however, that we will have to beat into the jungle ourselves, since no one has taken it yet.

One can follow Maxwell's way of thinking by assuming that all motions are directed as in mechanics, so that light and charges have to travel only one distance, and further assuming that only one type of electric charge of magnitude e, say the negative one, moves when an electric current flows. Then the electric current has a direction and can be modeled with a one-way description – exactly as in the case of light. But one can also pursue the goal of understanding the concept of opposites as a dualistic principle of nature, interpreting it as a holistic division or field branching process and applying it to the constitution of light as well. However, these are two different physical concepts that drive our understanding of nature and reality in opposite directions. And this is precisely "The Garden of Forking Paths" (Jorge Luis Borges). Anyway, in both cases that means:

Multiply Weber's constant by two and you get the speed of light, which is about 300,000 km/sec! Now you can probably imagine the surprise of Maxwell, which he first expressed in a letter to Faraday and in his essay "On Physical Lines of Force" in 1861, and then in the "Dynamical Theory of the Electromagnetic Field" of 1864:

"The velocity of propagation is the velocity v, found from experiments such as those of WEBER, which expresses the number of electrostatic units of electricity which are contained in one electromagnetic unit. This velocity is so nearly that of light, that it seems we have strong reason to conclude that light itself (including radiant heat, and other radiations if any) is an electro-magnetic disturbance in the form of waves propagated through the electromagnetic field."

Note that Maxwell has now changed his formulations: The electro-magnetic polarizations no longer propagate in a aether gas, but in an "electromagnetic" field – whatever this term may now mean, a certain tautology cannot be denied. To distinguish his interpretation from WEBER's original constant c, MAXWELL used the symbol V (for velocity). In 1905, EINSTEIN had still used V for the speed of light. A short time later, however, it was agreed to designate this constant as c again, without changing MAXWELL's interpretation.

By the way, I think that MAXWELL's interpretation of the nature of the motion of light, which seems quite convincing at first sight, is, next to the atomic hypothesis, the second deeply hidden assumption of which most physicists are not aware. This is exactly what leads to the seemingly insurmountable difficulties that physics has been struggling with for the last hundred years.

Interestingly, MAXWELL has never explicitly excluded a bidirectional electric current concept. Presumably because he was aware that his mathematical descriptions of polarizations using mechanical ideas of motion could not capture their true nature. From 1862 onwards, however, his way of thinking seems to have changed, since he now interpreted electric current as if it really had a preferred direction. The most important reason for this was certainly the intention to understand light and electricity with a uniform concept of motion. This could also have been the reason why he suddenly doubted the existence of positive electric charges and tended to regard them as missing negative charges. And if one leaves out half of the current model of AMPERE and WEBER, one obtains the model of a unidirectional electric current consisting of only one kind of charges.

Another reason might have been the exact determination of Ohmic resistance in telegraph lines, on which MAXWELL collaborated. In this context, it seemed quite natural to assume a directional motion of the electrical signals – after all, Morse code was transferred from A to B, just like light, except that light does not need a cable. But the transmission of electric signals was also explainable with a bidirectional electric current, as WEBER had already shown in 1857 with his *telegraph equation*. Although the theory assumed instantaneous actions between electric charges, the transmission of signals occurs at a fast but limited speed, which would be identical to the speed of light if the resistance of the wire were zero. This already indicates that it might be quite possible to dispense with a physical concept of time and still state a duration of structure-forming processes. In other words: Structural changes need time, but the determination of a duration depends on the size of the time interval for comparison. And this is relative, because it depends on the frame of reference – which is an interacting system. What sounds so cryptic at the moment will soon make sense. We already know something similar from special relativity: We always measure a constant c, which we interpret as constant speed of light, at least since MAXWELL. But for the light itself no time passes, so there are no speeds, and distances don't play any role either. Of course, we will come back to this interesting puzzle again and again in the next sections.

Summary:

WEBER's bidirectional electric current concept did not prevail, but essential ideas of WEBER entered MAXWELL's and LORENTZ's theory in an adapted form. MAXWELL's basic assumption that the propagation of light is a directed motion seemed to be in full agreement with all empirical facts and the measured speed of light in optical experiments. This assumption seems so logical and obvious that no one has seen a reason to doubt it so far. Moreover, the speed of light has been considered a law of nature since 1982, which also seems to confirm the correctness of the underlying assumption. However, the true nature (alias structure) of light has remained a deep mystery until today, as the unsolved quantum puzzle proves: It is about the seemingly unsolvable contradiction between the wave theory (divisibility of rays or waves) and the particle concept (energetical wholeness in the absorption event), which also applies to the true structure of matter. Quantum physics, which had actually originated to fathom this structure, cannot understand this structure, as long as the quantum is thought and treated as particle and the deficits of the wave theory are not understood.

Among these shortcomings is the definition of the speed of light, as proven by the EPR paradox and modern quantum experiments. So there are valid reasons to question it. If theoretical physics does not allow this any more, it puts a leg over itself. That the concept of a unidirectional propagation of light also contains a hypothesis about the nature of light becomes clear only when one realizes that light can branch. In optics, this is interpreted as a mechanical decomposition of wave bundles or light rays into their individual parts, i.e. not as branching in the physical sense. This is the great blind spot of 20th century physics.

A bifurcation process creates and changes structure of light, which challenges the continuum assumption of any field theory (with or without aether gas). But as soon as this phenomenon is treated as a structure- forming process, the concept of bidirectional motion of AMPERE and WEBER gets a new physical sense:

This type of motion can be understood as *simultaneous* propagation of two coherent branches in opposite (or different) directions, but also as a branched stationary field structure or standing wave consisting of two opposite energy forms interacting only during energy transmitting absorption events. Moreover, its now clear for us that such a field branching process must be reversible if the principle of conservation of energy is also to hold for point-like local absorption events in double-slit experiments. The same applies to the electric current pulse in the FARADAY induction experiment, which can be interpreted as a bidirectional branching and dipolarization process of an electric field. MAXWELL never explicitly explained what evidence led to his change of mind about WEBER's bidirectional electric current concept. But without experiments that allow a decision between the two theories of currents or to prove that light is a unidirectional motion – which would require a *one-way measurement* of the speed of light, which is *impossible on principle*, as EINSTEIN will explain to us later – a preferred direction of light and electric currents remains only a convention that obscures the true constitution of nature.

What makes the quantum puzzle so complicated is that MAXWELL's wave theory has been considered proven since HERTZ's experiments After all, they led to radio technologies whose success clearly proves that electromagnetic waves must exist in reality...

Hertz: Light magnified millions of times – the discovery of radio waves

Many contemporaries were deeply impressed by MAXWELL's design of an electromagnetic world view, but found the underlying physical conceptions quite difficult to comprehend, as HEINRICH HERTZ (1857-1894) reported in 1894:

"Many have eagerly set out to study Maxwell's work and, encountering no unusual mathematical difficulties, have nevertheless had to refrain from forming a fully consistent picture of Maxwell's views. I myself have fared no better. For all my admiration for the mathematical connections of Maxwell's theory, I was not always quite sure that I had guessed Maxwell's true opinion as to the physical meaning of his assertions. In my experiments, therefore, I could not orientate myself directly on Maxwell's book, I have orientated myself here on the work of Helmholtz, as can be seen from the presentation of the experiments."

MAXWELL's electromagnetic theory of gaseous matter in motion was by far not accepted during his lifetime (he died in 1879 at the age of 48 years). The field concept was not necessary in AMPERE's electrodynamics nor in the bidirectional current concept of WEBER. AMPERE had strictly denied the existence of fields, for they are neither tangible nor directly detectable. And so did many physicists at the end of the 19th century. They considered the existence of fields, the propagation of light in form of electric or magnetic polarization waves, and the prediction of other forms of light to be theoretical speculation. That the propagation of light in free space and even in the universe is basically the same process as dielectric polarization in ordinary insulator materials such as glass, ceramics, pitch and gases, which are somehow coupled with magnetic polarization processes, was a bold idea of FARADAY and MAXWELL that no one wanted to believe at first. However, between 1886 and 1888, HERTZ was able to prove experimentally that the waves of a different kind predicted by MAXWELL actually existed, and that these *"rays of electric force"* or *"waves of electric action"*, as HERTZ called them originally (called *radio waves* today) behaved exactly as visible light. The only difference was that they can be understood as light waves with very long wavelengths, as HERTZ put it: *"as if one were observing normal light magnified a million times"*. Thus our radios, televisions and mobile phones function by transmitting invisible light, to make the common clear for once.

With these experiments, HERTZ was able to prove MAXWELL'S assertions that both static and dynamic electricity can induce dielectric polarizations in insulator materials, which in turn produce the same electrodynamic (magnetic) forces, i.e. magnetic induction effects, as an electric current. This was basically proof that FARADAY'S induction principle applies universally to every kind of matter and even to free space. In other words, the world had to be electromagnetic. HERTZ found that extremely fast oscillations between opposite electrical polarization states produce radiation consisting of alternating "electrodynamic" (aka magnetic) polarization states that propagate in the form of *polarization waves* at the velocity of light. However, he was surprised to discover that these waves can *induce* electrical forces over very long distances, in contrast to the then prevailing theory that electric actions rapidly decrease with distance. He was able to confirm that these waves were 'electro-magnetic', that they penetrate walls, wood and dielectrics of all kinds, can be polarized in refraction, diffraction and reflection experiments, and can interfere like visible light. This discovery changed the scientific climate.

At this point, I have to insert one thing right away: One never really knows what the term *electromagnetic* is supposed to mean exactly. Somehow a clear distinction between electrical and magnetic properties of the waves seems to be extremely difficult. Even HERTZ had his problems with it. Practically, it just means that the electric and magnetic polarizations postulated by FARADAY and MAXWELL exist in reality and are just the two sides of the same coin, at least as far as the symmetry of the induction motion is concerned: Electric polarization makes *magnetic* polarization (an electric current or a voltage pulse induces a magnetic field, hence *electro > magnetic induction*); magnetic polarization makes *electric* polarization (a local changing magnetic field induces an electric polarization, aka electric current or voltage pulse, hence *magneto > electric induction*). This dualism does not really explain the difference – or commonality – between electricity and magnetism, since it describes only the effects. Yet we know they are pretty darn similar, as all mathematical models show since COULOMB. However, whether the opposite poles of a magnetic field actually correspond to opposite electric charges remains to be clarified. If so, magnetic and electric fields would be mirror symmetric manifestations of the same thing. So to understand HERTZ, MAXWELL, and the true nature of matter and light, one needs more cognitive clarity.

The ambiguities in HERTZ's descriptions can be traced to the conflation of the concept of electrodynamic force, which comes from AMPERE's theory of action at a distance, with the concept of magnetic fields and induction, which comes from MAXWELL's field theory and postulates near actions in a homogeneous substance. Then it becomes clear that 'electrodynamic' waves are identical with magnetic waves, as the following quotation from HERTZ himself proves:

"The hypothesis that the transverse waves of light are electrodynamic waves, already supported by many probabilistic reasons, is given a solid foundation by the proof that there are indeed electrodynamic transverse waves in air space and that they propagate at a speed equal to the speed of light." [29]

Clearly translated this would mean that light consists only of magnetic waves and fields. MAXWELL hinted at such an understanding several times; however, always remained ambiguous. If we adopt or introduce this understanding, it becomes clear that these waves can represent both temporal and spatial polarizations, that is, propagating magnetic waves and stationary branched magnetic field states, which becomes important for understanding quantum physics and the field branching hypothesis. So if HERTZ says that these waves are electrodynamic or represent electric forces, he could just as easily have said that they are *magnetic waves* which always induce electric polarizations. The whole question would be solved immediately, of course, if AMPERE's purely electrical theory of magnetism were correct. However, the question could be solved just as easily if electrical properties – and the structure formation of matter and light – could be attributed exclusively to magnetic fields. We will later see in the quantum physics chapters (volume two) that this is quite possible. Another problem is the third meaning of the term 'electrodynamics'.

29 Heinrich Hertz: "Über die Ausbreitung electrodynamischer Wirkungen", 4. Februar 1888. Reprinted in: Heinrich Hertz: "Untersuchungen über die Ausbreitung der elektrischen Kraft", Band II, J.A. Barth Leipzig 1891/94, S. 132. An English edition was published in 1893 under the title "Electric Waves"

It does not only refer to AMPERE and the connection between electric and electrodynamic (magnetic) actions, but also to the generation of electricity: If it is generated by moving charges, it is called electro-*dynamics*, in contrast to electro-*statics*. In my opinion, quantum physics raises the question how both phenomena can be brought to a common denominator. Consider that the motion of charges has been interpreted from the beginning as the mechanical motion of particles or volume elements. Since 1926, however, we know for sure that this idea cannot apply to reality, even if it seems to be didactically useful. In contrast, the experiments of quantum physics force us to understand this kind of motion as dynamic structural changes of stationary fields, which can only be interpreted as field branching processes and non-mechanical movements (as soon as one takes the superposition postulate physically seriously). This would be a new *kinetics* that combines electrodynamics and electrostatics and would also be applicable to molecular theory and a non-mechanical theory of heat. HERTZ was the first to pursue this connection theoretically and experimentally, proving that electrodynamics and electrostatics are indeed two sides of the same coin.

In 1884, just 27 years old, HERTZ begun his research with the question whether electrodynamic (aka polar magnetic) forces made by electric currents can make dielectric polarizations in insulators, and whether such electrostatic dipolarization processes can exert electrodynamic (polar magnetic) forces, as MAXWELL had claimed. Shortly thereafter he received tenure at the University of Karlsruhe (in Germany), where he began his experimental research on the known discovery that sparks jumped over the gap of an electrically charged copper rod exposed to high voltage by a RUHMKORFF spark inductor, which was one of the first *alternating current* transformers. The induction device generated strong electrical pulses with alternating electrical polarization of 100,000 volts from a direct current of only five volts. Sparks oscillated back and forth in the spark gap during discharge, so the electric current had no preferred direction. If the two ends of the rod were fitted with brass spheres at the gap, the sparks became even more visible because the spheres acted like electrostatic capacitors, which can store and release larger amounts of electric charge. HERTZ now accidentally discovered that the oscillating sparks in a second interrupted copper rod of the same design, which was lying a few meters away on a work bench, also induced sparks. He suspected some *kind of resonance* caused by invisible rays or waves emitted during spark discharge that could exert electrodynamic forces (magneto-electric inductions) over long distances, and began to investigate this phenomenon systematically.

In 1886, after optimizing his instruments, he was able to demonstrate this effect over a distance of nine and a half meters. He had thus practically built the first radio wave transmitter based on spark induction. Thirty years later, the further development of this principle would lead to extremely powerful spark generators and radio transmitters which could reach other continents (Germany's then most famous radio company was named Telefunken, which means Tele-Spark). To be able to detect the shape of the assumed electrical oscillations formed in space he used handy optimized, interrupted copper loops, on which sparks always jumped over when he came within the corresponding range. This hand- held wire loop device was one of the first radio wave receivers. With such a simple design he was able to investigate the behavior of these 'electrical waves' or rays more closely. By 1888, he had reached a distance of 15 meters between transmitter and receiver. And he was able to prove that this radiation actually existed and behaved exactly like light:

The sparks of the transmitter produced invisible radiation spreading out like MAXWELL's electromagnetic waves in all directions when no obstruction stood in the way. They could be reflected, focused and directed like light – only a half-cylindrical, parabolic metal mirror around the spark gap was needed. The radiation changed direction on a 600 kg hard-pitch prism – just as light is refracted on a glass prism. HERTZ proved the existence, the extension, the polarization, and the shape of the field with his hand-held wire-loop-resonator by walking around the hall comparing the strength and brightness of the sparks produced at various points. With this device he registered the local intensity of induction of dielectric polarizations in the copper loop, which allowed conclusions about the spatial distribution of the field strength. With this simple receiver, he was also able to demonstrate that the radiation is transmitted by the building wall, but can also be reflected when standing waves form. The wave was completely reflected by a large zinc sheet that he hung on the wall in another experiment. It causes the radiation to be mirrored, creating a standing wave in space between the transmitter and the zinc mirror that is invisible to the human eye. That this structure actually has a wave form, HERTZ could prove with his local intensity measurements: The invisible form had crests and troughs at which the receiver produced strong sparks (these are the points at which the electric polarizations that make up the wave reach their maximum), and wave nodes, where the handheld receiver registered nothing.

In this way, through reflection, HERTZ had created a large resonant oscillation, a *stationary* or *standing wave* with a wavelength of nine and a half meters (Fig. 25). The wave was a nearly plane vertical wave, because the spark gap was vertically oriented, and the wave height was about 75 cm. The wave reflected from the wall was in the same plane as the incident wave. Both waves overlap, interfere with each other and form a wave standing in space between emitter and reflector. HERTZ still assumed the existence of a magnetic wave (represented by the dotted line) oscillating at right angles to the electric wave in the horizontal plane, since MAXWELL's theory required that electromagnetic waves consist of electric and magnetic polarization waves that are at right angles to each other. The magnetic wave appears "phase-shifted" by a quarter wavelength and reaches its highest amplitude where the electric wave has its minimum (no spark induction at the nodes of the electric wave). However, since the wire loop can only detect magnetically induced electric induction, which manifests itself in the spark discharges, HERTZ adds the following remark:

"It should be noted that the determination of the position of the magnetic force here is made possible only by theory. From the experiments, the existence of a second kind of force besides the electric force cannot be concluded. [...] That this magnetic force produces effects, which cannot be explained by the electric force, will only [...later...] be substantiated by experiments, but only on the waves in wires."

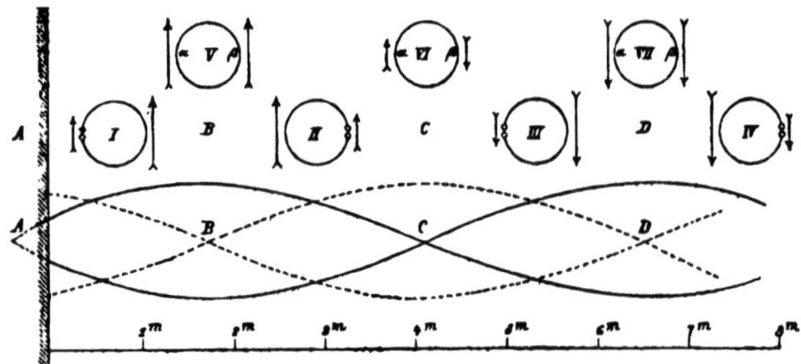

Figure 25 Hertz's illustration showing a 9 m long standing wave

Solid line: Incident and reflected electric wave emitted by a spark transmitter (from the right side, not shown). Left: Sandstone wall of the auditorium. The vertically arranged spark gap emitted a wave of vertical acting electric forces that reflected off the wall. The result was a standing wave nine and a half meters long with vertical electric polarization. Dashed line: Hertz also registered a horizontally acting force, which he interpreted as a magnetic wave according to Maxwell's theory. The circles symbolize the position of the spark gap of the ring-shaped hand receiver where the spark induction was most intense. The arrows symbolize the direction and strength of two electric force components. The oscillation period of the spark transmitter was estimated to be 155 million per second.

Both waves changes their force direction, or better, their polarization sign, at each wave node: When the electric wave oscillates in the vertical plane, the direction of electric polarity points downward in the wave troughs and upward in the wave crests. The same is true for the magnetic wave, which takes the same form in a horizontal plane, with the direction of magnetic polarization also changing at each wave node now between left and right. Thus, HERTZ believed to have proved that radiation waves always consist of two partial waves, one electric and one magnetic, whose wave planes are aligned at right angles to each other, as MAXWELL had predicted and illustrated, and that these waves are transverse waves (to the propagation direction of the beam). This perfect agreement, with both FRESNEL's and MAXWELL's theory, seemed almost uncanny. Has theory influenced the interpretation? (Fig. 26).

Could it not be that the type of measurement determines whether the wave appears electric or magnetic? This would not be quite the same as what MAXWELL illustrated. Moreover, there must be a crucial difference between traveling waves and standing or stationary waves. Today we know that this is the most important difference between light and matter, because atoms and molecules are understood in quantum physics as stationary wave formations, whose true constitution, however, remained unclear until today. With light, on the other hand, we assume that it always travels at the velocity c and that wavelengths and frequencies always play a role – even if light is imagined to consist of quantum particles. But how a particle can have a frequency remains unexplained. Understandably, this is also not understood to this day. Is this an indication that 'moving' light can turn into a *standing wave* by reflection? Does the reflection create a branched structure as in the case of a partial reflection or at a polarization filter? A standing wave, which then consists of two waves with opposite (orthogonal) polarization? Is the branched field then somehow rooted in matter? Can such a nodal or branching point be detected experimentally?

As far as I know, neither MAXWELL nor HERTZ considered that a light beam or wave could *physically* branch, for example, on a wall, an obstacle, a mirror, a semi-transparent mirror, a glass pane, or a polarizing filter. If this is the case, a certain interaction with matter must be the cause. This interaction must fulfill the role of an anchor point from which the partial waves or rays could propagate again at the speed of light. I think this could also take place when light interacts with dielectric materials (insulators). After all, HERTZ had already noticed something similar: electromagnetic waves cause dielectric polarization processes in non-conductive materials such as asphalt, paraffin, wood, paper, sulfur and glass blocks – just like the voltage applied to a dielectric capacitor in an electric circuit. And we know that they can store electricity as in the Leyden jar. However, they can also transmit, refract and decompose the electromagnetic waves like ordinary light beams at a glass prism, but can also reflect or re-emit the incident waves, as HERTZ noted. But if the incident wave is partially transmitted and partially reflected, it must branch

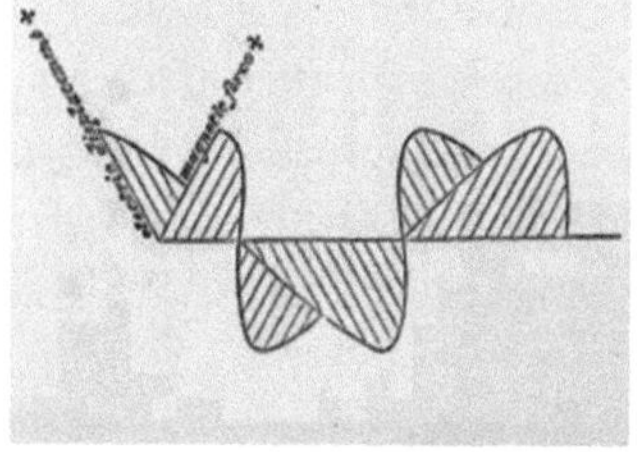

Figure 26 Maxwell's electromagnetic wave
Original illustration by Maxwell: One wave represents the "electric displacement" (electric wave), the other wave the "magnetic force" (magnetic wave). Parallel lines indicate the direction of the electric and magnetic polarization. According to theory, electric and magnetic waves always form an angle of 90°.

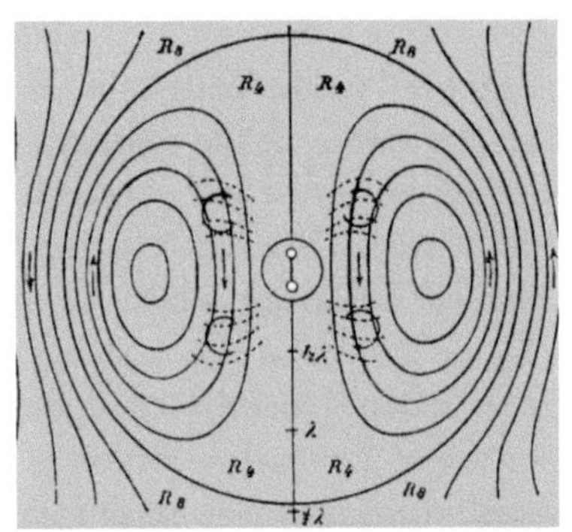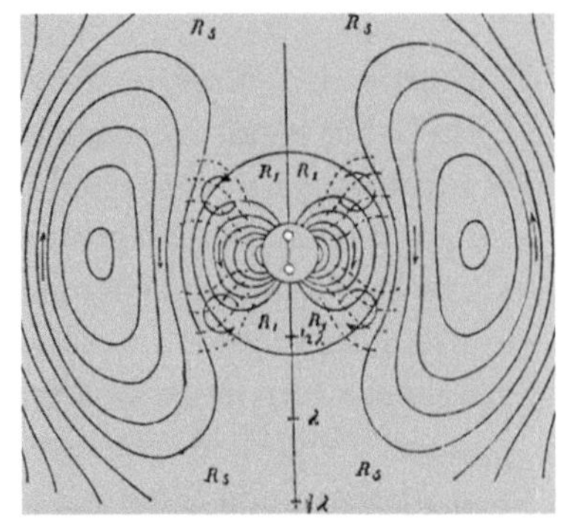

out. This would be the same as a *partial reflection* process known from semi-transparent mirrors and glass panes. So the re-emission could be exactly the process that describes the reflection of light if the matter does not absorb the incident light. This in turn would mean that matter absorbs light by dielectric polarization processes, which can only be structural changes in the integral molecules (atoms) and in the entire molecule ensemble. This is a hypothesis we should keep in mind until we get to the question of how to understand the emission and absorption of light quanta. This will also allow us to design an alternative to Bohr's hydrogen atom model, in which we replace the notion of particle-like electrons and orbits with dimagnetically polarizing branching processes. These can just as well be interpreted as dielectrically polarized branching states of a negative-electric field branch which has been further branched by the absorption of a light quantum.

So if the light in these experiments appears magnified millions of times, can we detect something that is normally hidden from our eyes? Can we directly observe and detect field branching processes? Hertz also looked a little closer and examined the radiation source. He found that it behaves almost like a gushing spring. Somehow electric and magnetic fields gush out of the sparking gap and form a bubble that constricts and divides mirror-symmetrically into two halves, which then detach symmetrically in the form of oscillating waves and spread out in space (Fig. 27).

When I saw these illustrations for the first time, I immediately thought of cell divisions and pulsating processes. If you know that the term 'electrodynamic' basically means 'magnetic', you can also see in these illustrations that the skipping spark generates transverse

172

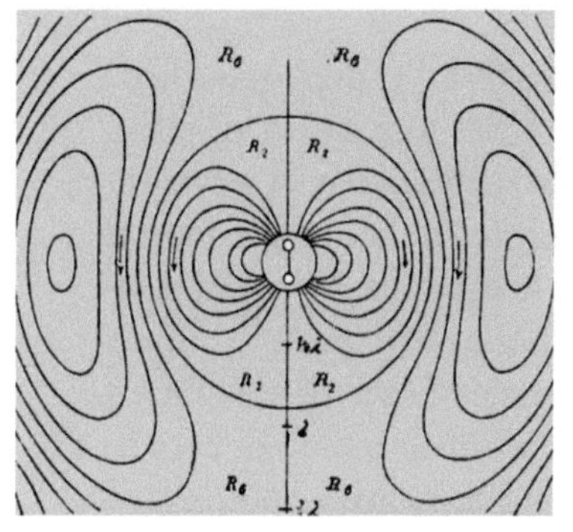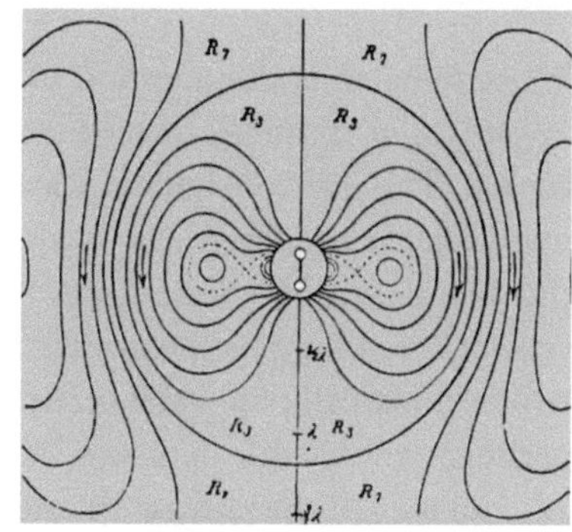

Figure 27 Hertz's field-line drawings from the radiation source

magnetic dipole waves (the spark gap, symbolized by the two white dots, is standing vertically in this experiment). Recall also that the detection is always done by magneto-electric induction, which also indicates that they must be magnetic waves. The mirror symmetry of the division process is striking; moreover, it's a dynamic process (the illustration shows four snapshots, to be read from left to right). However, it must be remembered that HERTZ illustrates his measurements with a topology of field lines that follows MAXWELL's theory and his own equations.

It soon became clear hat HERTZ had indeed produced a new kind of light, a previously unknown radiation of electromagnetic nature, just as MAXWELL had predicted. The experiments also showed that the spark gap represents an electric dipole, oscillating very rapidly between two opposite electric polarization states of the copper rods (depending on the electric current theory). This is an *electric dipole oscillator*. The discharges produce radiation of various wavelengths consisting of rapidly changing magnetic polarization states. And, as MAXWELL had also predicted, the experiments proved that such waves propagate through non-conductors such as air (and gases in general), glass, wood, ceramics, stone, brick, pitch, sulfur, etc. On reflection, this seems rather strange (or ingenious), because the light that propagates in our atmosphere and in the universe is then in principle the same as Maxwell's *displacement current* in the gap of an interrupted circuit, in a condenser, or in a block of raven-black asphalt! Since then there can be no doubt anymore that these waves exist in reality – if only there were not the tricky quantum puzzle!

What remains is the question of the waveform in the 'untouched', unmeasured or non-interacting state and the true nature of polarization. Do electromagnetic waves really oscillate flat in a two-dimensional plane, or is this just the result of the measurements, of the interactions? Can we discover in these experiments a confirmation of our idea that polarization is a branching process forming two ontologically opposite branches? Remember the double-slit experiment with single light rays or photons, which could not even be anticipated theoretically at that time: Logic demands a holistic division process, but in the case of light, the branched state cannot be observed or proven directly, because of energy conservation always both branches must contribute to the absorption event.

So every effective interaction consumes the energy of both branches (what special relativity cannot explain). This fact also makes clear why the amplitude of the (reunited) wave can only be measured as local field strength, which mathematically corresponds to the modulus of a squared quantity, which consists of opposites. In fact, the formalism states that the squared amplitude must consist of two opposite parts, which cannot be added in the usual way to model an energy transfer (like two halves consisting of the same substance), although it was clear that both must contribute to the result. You probably know something similar from your math class when the teacher points out that apples and pears cannot be added. These two parts are the roots, so to speak, that make up the actual polarization state. Thus, the two branches must be of opposite nature (what the words polarized and dipole originally mean). And because they are of opposite nature, in addition to our normal 'real' numbers, a new class of numbers must be introduced, the so-called 'imaginary' numbers and a different kind of summation, later called 'quantum mechanical summation of complex numbers'. We will return to this problem in the second volume, in connection with the interpretation that MAX BORN contributed to the Copenhagen Interpretation in 1927.

First we want to state that already in MAXWELL's formalism it was clear that the wave must consist of two hidden, not measurable energetic parts which enter quadratically into the determination of the field strength. So we can say that the wave which HERTZ generated and imaged with the help of his hand-held receiver is the result of measurements to which two parts must have contributed. The only question is whether these two parts are electric and magnetic waves, both contributing to the spark induction in the detector. Or whether the contour of the 'electric' wave, which Hertz has drawn with a solid line, rather describes the *resultant* of an acting magnetic wave, which consists of two orthogonal polarized magnetic partial waves, which only together can induce dielectric polarizations (aka electric currents) in the conductor. At the transmitter then magnetic dipole states would be generated, which according to MAXWELL show the dimagnetic division of an unknown gas, while according to the modern theory they are only excitations of an abstract field. This interpretation is so obvious that one wonders why HERTZ never considered it (this might have to do with his starting point, which was influenced by HELMHOLTZ' views). The fact is that this wave can only describe the *intensity* of local magneto-electric inductions (magnetic field strengths), which produce dielectric polarizations of various intensities in the detector, i.e. electric current pulses or voltages, which in turn produce the short-circuit sparks. It is therefore quite useful to make a clear distinction between magneto-electric and electro-magnetic induction when determining the nature of the field that produces these effects. In HERTZ's experiments, it is clearly magneto-electric induction. And the dielectric nature of voltage suggests that a bidirectional branching movement may well be considered if the true nature of electric current is to be understood.

Another problem is the motion of light. Is it really a motion in the sense of mechanics, a change of location? FARADAY and MAXWELL interpreted the wave as unidirectional motion of wandering polarization states. Is another view possible? But apart from these questions, the basic assumptions seemed to have been confirmed, at least for the time being, as HERTZ noted:

"The fact that the effect of induction propagates with finite velocity may already be considered as proven. [...] The most immediate conclusion is the confirmation of Faraday's view, according to which the electric forces are independently existing polarizations in space. For in the phe-

nomena examined by us, such forces are still present in space after the causes which produced them have disappeared again. So these forces are not only parts or attributes of their causes, but they correspond to changed states of space. The mathematical determinants of these states then justify that they are called polarizations, whatever the nature of these polarizations may be."

So Hertz could proclaim with full conviction: *"The wave theory of light is, humanly speaking, a certainty. [...] It is therefore also certain that all space of which we have knowledge is not empty, but filled with a substance capable of making waves, the aether."*
Finally, he also clarified that he should have better called his 'rays of electric force' or 'electric waves' electromagnetic waves from the outset: *...the expression 'electric force'...is only a name for a polarization state of space, to avoid misunderstandings I might have done better to replace it by another word, for instance the word electric field intensity..."*

Polarization and Branching

HERTZ's last remark brings us back to the crucial problem of how polarization is really to be understood and what kind of structural motion it represents. In the "Dynamical Theory of the Electromagnetic Field" of 1864 MAXWELL had already mentioned a statement of THOMSON, who had recognized *"that we must admit the existence of a motion in the medium depending on the magnetization, in addition to the vibratory motion which constitutes light"*. It was about the explanation of the FARADAY rotation of a polarized light beam by a magnetic field. MAXWELL thought that this missing motion could be a rotation around the axis of the magnetic field, which already points to the idea of 'spin'. This is the way quantum physics has taken sixty years later, but only with particles and probability waves. In contrast, we suppose that this missing motion is a dipolarizing branching process.

To examine the polarization of his waves, HERTZ used identical parabolic mirrors both for the transmitter and the receiver: When both standing vertically, sparks occurred in the receiver, but when the receiver was placed horizontally, no spark induction takes place.

Therefore the radiated wave had to have a certain polarization direction, in this case vertical, relative to the vertical aligned spark gap and the focal line of the parabolic emitter. This is what AMPERE has discovered and expressed with the relative positions of his 'current elements' in space, which was nothing else then a translation of the polarization rules from optics to electrodynamics.

To test whether the polarized wave could be further polarized like light, HERTZ built a large polarizing filter. This was a simple octagonal wooden frame with a diameter of two meters, in which one millimeter copper wires were stretched in parallel at regular intervals of 3 cm (the thing looks like a giant egg slicer). The octagonal shape allowed him to rotate the polfilter in 45° steps by simply placing it on the ground. When he placed the wire screen between the standing transmitter and the standing receiver, i.e. placed it in the vertical polarized wave, he found the following: If the wires are aligned vertically, i.e. parallel to the focal line, the supposedly vertical *electric wave* cannot pass through. But when the wires are aligned horizontally, the wave can pass trough! This was quite confusing because intuitively one would expect that at least a wave oscillating in the vertical plane should pass between the vertical wires without problems. But the experiment shows that vertical wires block the *entire beam*, while horizontal wires let the beam pass. Interestingly, many physicists and teachers claimed over decades that the beam can pass if the electric polarization plane is aligned parallel to the wires, as if they had never studied the work of HERTZ (Wikipedia, however, presents this correctly). The original text of HERTZ reads as follows:

"Now, if the two mirrors were set up with their focal lines parallel and the wire screen was inserted into the beam perpendicular to it so that the direction of the wires crossed the direction of the focal lines perpendicularly, the screen affected the secondary sparks almost not at all. But if the screen was placed against the beam in such a way that its wires were parallel to the focal lines, it intercepted the beam completely." [30]

30   Heinrich Hertz: "Untersuchungen über die Ausbreitung der elektrischen Kraft", Band II, J.A. Barth Leipzig 1891. Section 11: Über Strahlen elektrischer Kraft, Dec. 1888, p. 190 in the German book, in the English edition p. 177 (Electric Waves, 1893)

What does this mean for determining the plane of polarization of the incident wave, if one assumes that the beam consists of orthogonal polarized electric and magnetic waves at the same time? Suddenly it is no longer clear how the polarization of a light beam, which since FRESNEL has been interpreted as a wave oscillation in a certain plane, is to be understood – or where it is really located. HERTZ:

"The polarization of the beam is therefore not only that oscillations only occur in the vertical plane, but that the oscillations in the vertical plane are of electrical nature, while they are of magnetic nature in the horizontal plane. The question itself, in which of both planes in our beam the oscillation takes place, without indication, whether one asks for the electric or the magnetic oscillation, does not allow an answer."

This is a strange situation: FRESNEL's fundamental assumption that every light wave has a certain plane of polarization from the outset cannot be proven. Instead, HERTZ's further experiments show that polarizing filters bifurcate light waves, creating two partial waves with opposite polarization. So it is the "measurement", the inteeraction with matter, which produces the polarization. We have already discovered this in FRESNEL's wave theory: A wave can branch into two partial waves of opposite nature. Their oppositeness is expressed in orthogonal polarization states, which appear as exact right angles in our three-dimensional space. Amazing (some people claim to this day that nature does not know right angles). And since FARADAY we know that this must be a magnetic polarization – consequently the "electric" wave of HERTZ must have been in reality a magnetic wave which is now branching at the polarizer!

HERTZ created such a bifurcation by rotating the wire polarizer by 45°: If the angle of the wire screen polarizer is plus or minus 45° (relative to the focal line), the incident beam branches into two partial rays: One passes through, the other is reflected. They are now polarized at right angles to each other. HERTZ was able to prove this with his parabolic receiver, which he first placed behind the wire screen. If it produced sparks in the vertical position behind the screen, the same receiver, placed in front of the screen, produced sparks only in the horizontal position. The incident beam must therefore have branched in two orthogonally polarized rays. In this way it becomes clear that the 45° wire-grid polfilter causes a spatial separation, a branching, into two partial rays with orthogonal polarization. This proves that polarization is a relative property between two opposite parts which cannot have arisen by an ordinary splitting and doubling. Why this oppositeness appears in this case as orthogonality of the polarization directions and not as anti-parallelism of the planes (which would also have been conceivable) remains to be clarified. So are the electric and magnetic properties of a light wave also a product of measurement? As HERTZ noted, the distinction between electrical and magnetic properties actually depends on the type of measurement:

"The experiments were concerned only with the propagation of the electric force. It was desirable to show that the magnetic force also propagates with finite speed. According to the theory, however, the generation of special magnetic waves was not necessary for this, because the electric waves would have to be waves of the magnetic force at the same time; it was only important to really prove the magnetic force in these waves besides the electric force."

HERTZ tested this in another experiment with an aluminum ring in which the wave created magnetic forces that rotated the ring. This quotation also shows that he was aware that there is no real difference between electric and magnetic waves. And, also importantly, HERTZ concluded by *theoretical reasoning* that his waves must be traveling at the velocity of light. He never measured the speed of the waves experimentally; he calculated it from MAXWELL's theory: The electric dipole oscillator frequency of the transmitter was estimated, while only the wavelength between the nodes of the standing wave was measured. However, these wire polarizer experiments showed the same as with visible light. As HERTZ put it:

"The screen behaves to our beam exactly as a tourmaline plate behaves to a linearly polarized optical beam. Apparently, the screen decomposes the incoming wave into two components and lets through only the component that is perpendicular to the direction of its wires."

Isn't that a nice circumscription of a branching process? The same results were already known from partial reflection in optics. Hertz has clearly shown that the branching process of radio waves at the 45° wire-grid polarizer is the same as that of normal light in partial reflection experiments, only that the wavelengths here are millions of times larger.

This shows that branching processes cannot depend on the wavelength or intensity (frequency and energy content) of the radiation. Thus, the ability to branch must be a universal, scale-independent property of electromagnetic radiation. We also know that the two rays must still be physically connected after the division, otherwise they could not interfere. This is exactly what *coherence* means physically in its original sense. And we know that the 45° wire polarizer experiment is equivalent to the experiment of partial reflection in optics, where a light beam hits a glass plate at an angle of 45° and branches: One partial beam passes through, the other is reflected. From all this follow the most interesting questions of physics: How do they act together in the case of a quantum absorption event?

In partial reflection experiments we will find the answer. Unfortunately, Hertz did not made experiments with a 45° polarizer and two receivers simultaneously. He probably would have carried out these experiments later, but he died suddenly in 1894, shortly before his 37th birthday. So perhaps we should catch up with Hertz's experiment in his honor, but this time with two receivers set up in front of and behind the 45° wire grid. Then we would be able to study in detail how the field structure is constituted on a much larger scale than Maxwell could ever have expected: We could explore the structure of light as if through a giant microscope with millions of times magnification (compared to normal visible light). Imagine: We were able to walk around directly in an experimental setup that normally takes place on lab benches and hides submicroscopic processes! And we could find out if the radiation behaves holistically in branching processes and quantum absorption events. This would be an amazing art installation (provided, of course, that the intensity of the electromagnetic radiation is harmless). Unfortunately, since Hertz passed away much too early, such questions did not surface until decades later in quantum theory.

What else Hertz found out about the nature of electric current
Metals are opaque to light and electromagnetic waves of all wavelengths. He found this out with a coaxial conductor, which he thus practically invented: The conducting wire is encased in a metal cage that shields it from external magnetic and electric fields. As HERTZ was able to prove experimentally, the incident electromagnetic wave can neither penetrate metal surfaces nor metal grids, as long as the wavelength is greater than the grid width. Consequently, the wave must form wave nodes at the interface, i.e. points, lines or surfaces where the electric and magnetic field strength becomes zero. These are anchor points and surfaces at which the wave is mirrored and thus, in a sense, fixed. A metal wire or rod practically forms a sink in space to which electromagnetic fields literally stick because they are reflected. At these reflection points, the polarization of the wave instantly changes to the opposite state, which means that the polarization state is mirrored.

HERTZ was sure to have proved that radio waves need time to propagate. The concept was that these waves describe cyclic processes in which the aether gas oscillates harmonically between two opposite polarization states. The polarizations are generated locally, but propagate spatially in the gas. This seemed to prove the near-action concept as well. However, the propagation in the aether gas should not change the original oscillation rate (frequency). This is not the case in ordinary gases, since the excitation energy decreases accordingly as it is distributed over more and more matter. Electromagnetic waves, however, should be able to produce electrical induction effects at the original frequency even at very distant points. Since the advent of quantum physics, however, we know that this wave conception, while still needed, cannot capture the true structure of light. Spin phenomena and EPR paradoxes have compounded the problem. So there must still be significant gaps in our understanding. We already notice one at this point: It is unclear whether one can speak of a mechanical motion (change of location) and thus of a speed of light at all for standing light waves.

So, from HERTZ we have learned that electromagnetic waves cannot penetrate metals. From FARADAY's diamagnetism we know that magnetic fields cannot penetrate metals such as copper because the incident magnetic field is counteracted by a shielding antimagnetic field in the metal rod. And from FARADAY induction we know that a changing or approaching magnetic field makes an electric field (a dielectric polarization), which can be measured as a voltage spike or current pulse. But if neither electric nor magnetic fields can penetrate metals, how do voltage and electric current in the conductor come about? Moreover, HERTZ found that the electric current cannot consist of moving negative particles in the conductor, because they would move very slowly, practically at a snail's pace. It seemed almost as if they could only oscillate locally. It was therefore not possible to explain the rapid transmission of electrical signals with moving negative charges in the wire, as MAXWELL had originally hoped. Apparently, OERSTED was right after all with his assumption that electric charges move along the outside of the conductor. This result prompted HERTZ to make the cryptic remark:

"Undoubtedly, metals are non-conductors of the electric force, for this very reason they force it under certain conditions not to disperse but to remain together, and thus become conductors of the apparent origin of these forces, of electricity, to which the usual terminology refers."

According to HERTZ, one should distinguish between true and free electricity: True electricity is bound in and between molecules. This structrual electricity of matter cannot be altered by electrodynamic processes, i.e. electric currents. However, the electric polarizability of molecules belongs to free electricity. With regard to free electricity, even dielectrics can be considered conductors. This apparently means that almost all matter can be dielectrically polarized by magnetic induction, while the conductivity of metals must be related to their *reflective* structural properties. Di(a)magnetism must also play a role, since a mirror-symmetric magnetic field structure is formed in and around the conductor (incident magnetic field outside, anti-field inside). HEAVISIDE must have thought along similar lines: In 1892, he claimed that the electric current is caused solely by external magnetic fields. This did not contradict the experimental findings, but changed the perspective on causality: until then, it was believed that the electric current makes the magnetic field around the conductor. HEAVISIDE countered dryly: *"We reverse that, the current in the wire is produced by the energy transmitted through the medium surrounding it."* If one thinks this through, then the magnetic field is the primary cause of electric currents and dielectric polarizations.

Tesla: Radio pioneers, power stations and undaunted inventors

The research of HERTZ had convinced even the last doubters, both of MAXWELL's theory and the existence of aether waves, although many questions remained to be cleared. HERTZ's experiments rightly went down in history and inspired his contemporaries enormously. But the new electromagnetic universe still seemed to be full of mysteries, and many researchers were underway to explore it. Only thirteen years later, the first radio waves crossed the Atlantic, something HERTZ had still be considered completely impossible.

As MAXWELL had already suspected and HERTZ had shown, visible light was a special phenomenon only insofar as man seemed to be tuned to this wavelength. This radiation was directly visible to the eye, while infrared radiation was directly perceptible to the skin as heat. Radiations of other kind were thus basically also only light, although not perceptible without technical aids. The universe could then be full of invisible radiation, or full of potential energy.

OLIVER HEAVISIDE had shown in his *Electrical Papers* (1892) that the Earth's crust indeed behaves as a dielectrically polarizable material that acts like a good conductor, which led to the idea of antennas in the form of a half-dipole. This means that only one rod of the Hertz dipole oscillator is needed as an antenna when electrically connected to the ground. The second part of the electric circuit is the Earth's crust, which behaves like a huge dielectric block. This configuration would significantly improve the range of radio wave transmissions, since the ground waves do not propagate three-dimensionally like light waves in air and space, but generate two-dimensional dielectric polarization waves and stationary states like in a glass block (compare Fig. 22, p.135). This would mean that the energy radiated through the Earth's crust would not decrease with the square of the distance, but only proportional to the distance, while the waves radiated by the antenna and transmitted through the air would have to be dipolarized magnetic waves. This conception was based on a clear separation between dielectric polarization waves and states in matter and dimagnetic polarization waves in free space, as postulated by MAXWELL, although somewhat ambiguous.

NIKOLA TESLA (1856-1943) was one of the first to think about how to use alternating current to transmit electric energy over long distances and to drive electric motors. He also thought about how to harness the potential electric energy stored in dielectric matter. As early as 1882, years before HERTZ proved the existence of 'electric' waves, he invented the rotating magnetic field and several types of alternating current motors. Between 1884 and 1889, now living in the USA, he invented, at a breathtaking pace, multiphase alternating current systems, two- phase, three- phase and multiphase electric motors and generators, high-voltage transformers and novel arc lamps that operated on the basis of wireless induction transmission. In 1887 he begun a cooperation with the WESTINGHOUSE to establish his alternating current (AC) electric power system on industrial scale. EDISON saw his direct current (DC) monopoly endangered and unleashed an economic war, combined with a dirty press campaign against TESLA and the WESTINGHOUSE corporation. But even EDISON could not prevent technological progress. As early as 1894 – the year of HERTZ's death – the (then) most powerful hydroelectric power plant of the world was built at Niagara Falls, equipped with three huge two-phase TESLA generators of 1,500 horsepower each, which went into full operation in 1896. TESLA's invention and development of the AC system was so groundbreaking and significant that this technology still forms the basis of electrical power supply in all industrial countries today. But TESLA's greatest vision could not be realized to this day: The limitless, extremely inexpensive generation of electrical energy and its wireless transmission over arbitrarily large distances around the globe.

Between 1887 (the same time as HERTZ published his first experiments) and 1891, TESLA applied for about 40 patents on AC motors, generators and circuits, covering the whole AC technology. Further patents followed on high frequency current generators, transformers and condensers, antennas, the grounding principle, and wireless transmission. During this time, he invented the principles of transmitting long radio waves by generating high-frequency electrostatic oscillations in specially designed capacitors (containers of dielectric materials that had a volume of about one cubic meter) and converting them into powerful electromagnetic waves with his Tesla transformers. The basic intention, however, was not to emit radio waves solely for the purpose of message transmission and communication – that was a useful side effect for TESLA, but not the main thing. His basic idea was to suppress the lost of radiation energy through the antenna, because emitted light, i.e. electromagnetic radiation, like thermal radiation, cannot be recovered. TESLA's real intention was to impress electrical tensions, i.e., dielectric polarizations, onto the giant insulator block called Earth. Since the Earth's crust is a dielectric, it should be possible to store huge amounts of electrical energy in it, similar to the glass body of a Leyden Jar. The uniform distribution of electric energy in form of dielectric polarizations of molecular structures in the earth rocks and soil layers would make it possible to tap this potential energy locally, wireless, and with minimal transmission losses at any point on the surface with suitable receivers. Because we remember, even according to the findings of HERTZ, dielectrics are virtually transparent to long electromagnetic waves and thus good conductors. Just remember the asphalt prism and the sandstone walls in the experiments of HERTZ, which were transparent for radio waves like glass bodies. So the only question is how thick such a block can be, what wavelength is optimal, and how the ratio between stored and transmitted energy is. This would tie directly into KOHLRAUSCH's basic electrostatic research.

To many electrical engineers, radio technicians and physicists, however, TESLA's visionary ideas seemed weird and fanciful. This does not seem to have changed to this day. But basically these are ideas in full accordance with FARADAY's discoveries, MAXWELL's field theory, and HERTZ' findings. Exactly the same problem was also the starting point for HERTZ in 1887, before he discovered the propagation of radio waves through space: To prove that electrodynamic forces magnetic fields) can induce dielectric polarizations in insulator materials like sandstone, wood, paper, glass, ceramics, paraffin, asphalt, sulfur, gases, petroleum, etc. And counter-wise, that dielectric polarizations of insulators could induce electrodynamic forces (aka magneto-electric induction actions) far away.

TESLA's concept of wireless power transmission would still be a revolutionary technology today: Imagine what that would mean for industrial and residential power, the storage of solar and wind energy, heating technologies and the electric car industry: Availability of power and charging stations anywhere, independent of wired power grids, providing power where it is needed. Vehicle batteries, which are too big, too heavy, and too expensive, will become obsolete or at least significantly smaller; the same goes for stationary battery storage. In principle, a simple diode detector would suffice, requiring neither a battery nor a power supply for its operation, but only an antenna and a ground connection. The incident radio waves induce alternating currents in the antenna, which are converted into *direct current* pulses by the diode. This is exactly how the simplest radio receivers work, which are equipped with simple crystal detectors and headphones and do not require a power source at all. However, in a "Tesla detector" the main energy would be transmitted through the ground connection and not through the antenna. In this way, electricity could be generated anywhere and sourced from anywhere. Is this really physically impossible, as many physicists seem to believe? If I have understood correctly, at least in principle, what FARADAY, MAXWELL, HERTZ and TESLA discovered and thought up already in the 19th century, TESLA's ideas, inventions and developments are purposeful extensions and creative applications of both the electromagnetic field theory and electrical engineering. These only went far beyond the current state of knowledge and technology. So the only question is how deep these invisible light waves can penetrate into the ground, what distances can be bridged, and whether there are any dangers to humans and the environment.

TESLA's ideas were bold and visionary, but always those of a practitioner who was also a great scientist and entrepreneur. Yet another inventor and pioneer named GUGLIELMO MARCONI, in 1896 just 24 years old, had another commercial application in mind. He developed his own wireless technology in Italy between 1896 and 1900, then in England with generous financial support from the British

Postmaster General. In 1900, he succeeded in building and bringing to market the first working wireless radio system. Yet, parallel to the dynamic invention and technology development processes in industry, strange academic prejudices spread at some universities.

HERTZ believed until his death that radio waves were of no practical use. And like HERTZ, many physicists considered the wireless transmission of energy over long distances to be impossible on principle. HERTZ firmly believed that transmitters, receivers and devices of huge proportions would be needed, drawing this conclusion from the analogy with optics: Since the wavelengths of radio waves were millions of times greater than those of visible light, theoretically all devices would have to be magnified on the same scale (he was probably thinking of his 600 kg asphalt prism). Moreover, the light emitted in straight beams would have to escape into space after three hundred kilometers because of the curvature of the Earth, at least according to the prevailing theory at the time.

But all these experts were wrong, as MARCONI was able to prove in 1901 with the first radio wave transmission across the Atlantic over 3,500 km. This shows that nature always hides secrets that cannot simply be deduced from equations and the currently valid physical theories. And this also proves that even successful experts are able to prevent scientific and technical progress in good faith. Incidentally, the problem of wave theory was solved by HEAVISIDE in 1902 with his "ionosphere" hypothesis, in which he postulated a layer in the upper atmosphere consisting of negative charged molecule ions that reflected radio waves. TESLA pursued a similar concept, which played an important role for his wireless power transmission. However, most experts did not share these views, because according to the then current physical theory, this would have meant that waves propagate at a speed faster than light (due to a phenomenon known as total internal reflection). Only later it became clear that the ionosphere hypothesis was correct, which forced physicists to distinguish between phase and amplitude velocities of light. And the fact that the concept of a velocity no longer makes physical sense in terms of standing waves created by total reflection was not well understood until the year 2000, when physicists began talking about 'standing light' confined in crystals.

In 1943, after decades of ligation, the U.S. Supreme Court declared TESLA to be the inventor of wireless radio technology. As it turned out, MARCONI's radio technology was not capable of covering such distances. According to the Supreme Court, he had quietly exploited TESLA's patents, copied TESLA's oscillators, transformers and circuit diagrams. This was the only way he could bridge the Atlantic. TESLA had sincerely congratulated Marconi on his success in 1901, but when he later learned of the facts, he was simply disappointed in MARCONI's character. Carelessly, TESLA had generously refrained from legal patent disputes, because at that time he was still earning a fortune from his Westinghouse patents, which he immediately reinvested in his research laboratories and machine developments. Once WESTINGHOUSE got into serious financial trouble, TESLA just as generously waived his royalty income from the AC patents forever, which got him into real trouble just a few years later. By 1902, he could no longer afford his expensive research and development activities. Moreover, he was unable to complete the Wardenclyffe project on Long Island, which was to launch his wireless power and information transmission company. Incidentally, that the principle of wireless energy transmission actually works, he had successfully demonstrated in a one-year field trial in Colorado in 1899.

This brief history of a new emerging technology illustrates that the relation between science and technology is neither direct nor linear. Expert *opinions* on the significance of inventions that have not yet been made, applications that have not yet been developed, and technologies that have not yet been experimentally explored should be viewed with reservation. This is because the truly new is, by its very nature, unprecedented. Fortunately, most inventors at that time did not believe, as many scientists, engineers and business managers do today, that inventions, technologies, progress and the nature of reality can be derived directly from the theories that are just fashionable – and continued to experiment undaunted.

III. How the electromagnetic world view failed, for now
In search for the electromagnetic structure of matter

Today's physics is not far away from the question if not
everything that exists was created from the aether gas.
(Heinrich Hertz, Heidelberg 1889)

On the way to a comprehensive electromagnetic world view?

Although the industrial use of electricity was already in full swing at that time, theorists apparently did not yet fully understand how electricity actually works in detail and what electricity and electric current really is. As HERTZ made clear, there were still many open questions and much things to discover. In 1889 in Heidelberg, Germany, he outlined the expectations on a comprehensive electromagnetic world view of physics with the following words, which sound highly topical again today:

"There is before us the question of the unmediated remote effects in general. Do such effects exist? Of many that we thought we had, only one remains – gravity. Does this also deceive us? The law by which it works already makes it suspect. In another direction, the question of the nature of electricity is not far off. Seen from here, it is hidden behind the more specific question about the nature of the electric and magnetic forces in space. Immediately following this, there is the enormous main question of the essence, the properties of the space-filling medium, the aether, its structure... It seems more and more as if this question overrides all others, as if the knowledge of aether must reveal to us not only the essence of the former imponderables, but also the essence of matter itself and its innermost properties, gravity and inertia."

After the experiments of HERTZ, MAXWELL's electromagnetic field theory was immediately taken seriously, even in continental Europe. This led to WEBER's concept of bidirectional electric current being rejected too quickly and never being able to develop into an alternative bidirectional field theory. MAX PLANCK, then an upcoming star of theoretical physics, was not entirely uninvolved in this. In 1887, just appointed professor in Kiel, he renounced an adequate appreciation of WEBER's experimental and theoretical contributions to the field of electrodynamics. But this had nothing to do with a professional discussion of scientific arguments or different view points. PLANCK did this in deference to HELMHOLTZ, who was his professor in Berlin, as he was also for HERTZ some years ago. At that time HELMHOLZ and WEBER had a serious scientific argument about the conservation of energy, which HELMHOLTZ ended rather abruptly. The emerging tendency to use MAYER's energy conservation principle (which HELMHOLTZ had adapted and promoted in physics) as a shortcut for overly complicated theoretical considerations – and to skip overly diffuse physical concepts – was probably also the reason for HERTZ to discard some of the physical concepts and equations of MAXWELL. HERTZ believed that the idea that aether molecules are magnetically polarizable like ordinary gas molecules was superfluous, although he believed in the aether gas. In addition, MAXWELL's mathematical notation seemed to him to be rather cumbersome. So he adapted the MAXWELL equations and reduced their number from twenty to four. A similar reduction was proposed by HEAVISIDE who introduced the new form of vector notation. And later some additions were made by ANTOON LORENTZ, who reintroduced important ideas of WEBER's electrodynamics into the electromagnetic field theory like the acceleration of charges and the speed dependence of electric and magnetic forces – which in turn MAXWELL had considered superfluous.

MAXWELL's electromagnetic field theory was only really recognized after the experiments of HERTZ, which led to many new inventions, ideas and technical applications in the emerging electrical, telegraph and radio industry. Between 1888 and 1898 all signs indicated that the world could indeed be explained by a unified electromagnetic theory. Nobody questioned the assumption of the aether gas, for it seemed self-evident that wavelike propagating polarization processes can only occur in matter. Towards the end of the 19th century, many physicists were convinced that physics was heading towards a comprehensive electromagnetic world view. The new world view was supposed to be able to replace the atomic assumption and the mechanical world view, and to derive Newtonian mechanics – the change of location of material bodies and gravitation – from purely electromagnetic forces and principles. But for this, a decisive ingredient was still missing: The design of matter models which should be able to represent the constitution of molecules either as local-spatial condensations of fields or as structural formations of an as yet unknown primordial gas with extremely low density, from which all elements of the periodic table would have to emerge with increasing density and mass. The idea that there could be holistic division processes of gaseous matter and light causing such a structure formation was still not born.

Of course, since 1805 it was known that NEWTON's atomistic particle hypothesis of light had to be abandoned because of the division process in double-slit and interference experiments, what led to the wave theory of light. But it was simply not recognized that the atom hypothesis could only be saved by the postulated preconfiguration of the molecule and a mechanical splitting of the same, which was experimentally not derivable. In this way, the atom idea survived in chemistry and physics, although almost everyone knew at this time that the idea of indivisible tiny billiard balls was only a preliminary, albeit very useful abstraction.

The debate about whether atoms actually exist physically continued among physicists and chemists until 1910. After that, the question seemed to be settled once and for all – until quantum physics made the problem even more acute fifteen years later. More than a hundred years had passed since AVOGADRO's molecule hypothesis until it was recognized in 1926 that electrons, atoms and molecules, just like light, had to be treated with wave theory because they had to be divisible, *at least theoretically.* The idea that these could be holistic division and branching processes was not exactly obvious, although quantum physics was already so close to the essence of these divisions that it really should have stumbled upon this discovery. This was mainly due to the sudden renunciation of deeper insight, which leading quantum physicists as BOHR and HEISENBERG voluntarily imposed on themselves in 1927. At that time, everything already indicated that the atom hypothesis, i.e., the philosophical assumption of indivisibility, had failed. But no physicist understood this, since the word 'atom' in common usage had come to be understood as a synonym for mass particles. So quantum physicists asked themselves in bewilderment: How can it be that atoms and elementary particles cannot really exist, if molecules and mater really exist? Not understanding the problem at all, they thought they had to save the atomic and elementary particle hypothesis as a concept 'useful' to physics – and built around it an interpretation of quantum 'theory' that made no physical sense at all. Interestingly, as early as 1870 MAXWELL was well aware that the atomic paradigm blocked a deeper understanding of the true nature of matter:

"The ancient theory that matter is composed of atoms ran into a logical paradox. On the one hand, atoms were supposed to be hard, impenetrable and indestructible. On the other hand, the evidence of spectroscopy and chemistry showed that atoms have internal structure and are influenced by outside forces. This paradox had for many years blocked progress in the understanding of the nature of matter".

Today we have to admit that this blockade of cognitive progress still exists and is not understood in the least: This is the apparent contradiction between *wholeness* and *divisibility*. That is the actual natural philosophical and physical problem that hides behind the quantum puzzle, the wave-particle paradox, which has not been solved until today (physicists who claim that the problem was already completely solved in 1927 have not yet understood the real problem – so do not believe them and form your own judgment). As FEYNMAN put it once: *„Science is the belief in the ignorance of experts."*

As MAXWELL had made clear with his theory, there can be no essential difference in the constitution of matter and light. Both consist of electrically and magnetically polarizable structures. Since it was not clear how the term *structure* could be understood and defined in a non-mechanical way, MAXWELL and his successors could only approach the structure of matter and light with auxiliary ideas from mechanics combined with field descriptions. But if one forgets that the physical conceptions determine the mathematical model design, the apparent precision only disguises the weaknesses of the underlying conceptions. This is exactly what became the hallmark of physics in the 20th century.

The discovery of new radiations

At the beginning of the 20th century, it was not yet clear how matter interacted with light and how the structures involved changed as a result of the interaction. But, as MAXWELL had predicted thirty-five years earlier, new radiations had been discovered, while spectroscopy was about to provide crucial clues about the structure of matter. But with these new discoveries, the problem of understanding the unity of matter and light seemed at first to become even more complicated:

The first complication resulted from the violation of the *symmetry of motion*, which appeared in two seemingly different situations: First, in the question why there is no speed difference between the two branched light beams of an interferometer, although the interferometer moves with the Earth through the aether. This problem found its expression in the MICHELSON-MORLEY experiment. And second, in the understanding of electromagnetic induction, where it plays no role if the magnet or the wire is moved. In both cases it is about understanding the symmetry of the polarization process itself and the symmetry between electric and magnetic polarizations. But even MAXWELL did not succeed in making it clear how the structural similarity of matter and light can exactly be understood, and how the gaseous aether – or the electromagnetic field – exactly interacts with ordinary gaseous, fluid and solid matter. Originally, the question was how the phenomenon of polarization can be understood physically in order to be able to explain a similar structure formation of matter and light. Based on this physical principle, it should then be possible to model the electromagnetic properties of nature mathematically and geometrically with field models – provided, of course, that polarization was physically correctly understood. The fact that MAXWELL had not succeeded in this did no longer seemed so important after HERTZ's experiments. However, the mechanical auxiliary terms that MAXWELL had introduced to model the phenomenon of polarization obscured the fact that the term polarization actually means the division of a whole into two opposite parts. So the kind of motion MAXWELL could not imagine was a *non-mechanical* bidirectional motion, that is, a holistic division, bifurcation or branching process. Therefore, auxiliary concepts based on unidirectional motion cannot lead to true structure conceptions.

The second complication arose from the discovery of X-rays in 1895 and of radioactivity of radium and uranium in 1896. In 1898, it was found that radioactivity consists of so-called alpha, beta and gamma rays, presumably produced by the decay of atoms or molecules. At that time, the existence of atoms had not yet been proven, but with the discovery of new types of radiation, the so-called atoms and the molecular structure could be studied in greater detail, creating a whole new field of research. Some of these rays seemed to behave like light, others like gases and longitudinal impulse waves.

• *X-rays* behaved like electromagnetic waves or light with extraordinary high energy manifested in a very high frequency and extremely short wavelengths.

• *Alpha rays* turned out to be *helium gas* rays, which were electrically positive polarized and behaved like longitudinal waves. So these elements had to contain helium, or to produce it in the decay process. Once again, gas theory and the problem of polarization came into play. In the kinetic theory of gases, which MAXWELL had improved significantly (but had no connection to his aether gas hypothesis), it was imagined that gases consisted of individual, space-like separated particle-molecules. Therefore, it was imagined that the emitted gas jet consisted of particle-like molecular ions with a certain speed. In this way, the idea of "particle rays" and "matter radiation" was born.

• *Beta rays* turned out to be negatively charged rays, which shortly thereafter were interpreted as electron particle rays.

• *Gamma rays* turned out to be high-energy electromagnetic waves with less than energy than X-rays, but more than ultraviolet light.

The third complication resulted from a discovery that HERTZ' assistant HALLWACHS had already made in 1886: This was the light-electric effect, also called *photoelectric effect*. He found out that the surface of a metal plate becomes negatively charged when the incident light beam has at least as high an energy as ultraviolet light. He also discovered that UV-light cannot penetrate window glass, but quartz glass can. Around 1900, experiments by LENARD, also a former student and assistant of Hertz, proved that light-induced electricity does not depend on the intensity (brightness) of the incident light beam, as it should according to Maxwell's wave theory. That was of utmost importance: The effect does not depend on the amount of incident light rays, but only on their frequency, which also defines the color of the light. The higher the frequency – or the smaller the wavelength – the higher the energy content of the light wave. That means that light with a high frequency is needed to electrostatically charge a metallic surface. Thus, even a weak ultraviolet light beam (consisting of a few rays, or even just one light ray) can trigger the photoelectric effect, but intense yellow light cannot.

This raised the question of how the electrical charging of the metal surface occurs. Do (magnetic) light waves induce dielectric polarizations? Can UV-light polarize metals and create opposite electrostatic charges that can be diverted as 'photoelectric' current? Wasn't that exactly the same as Faraday induction, where a magnet approaches an metallic conductor and induces an electric current pulse? Or was it just reflected light that loses its neutrality through interaction with molecules in the boundary layer and appears as 'negatively charged' light? Some physicists believed that light-induced electric rays were negatively charged light rays related to ultraviolet light. As it turned out, however, the reflected negative rays behave exactly like *cathode rays*, which Lenard had already studied in the years before.

What was the nature of the cathode rays? They were originally discovered by Faraday and further researched by Crookes, who presented the state of research in a review lecture to the Royal Society in 1876. When a high electrical voltage is applied to a vacuum tube containing a highly diluted gas, the gas begins to glow. The glow is caused by either negatively charged molecule ions or electric charges emitted from the heated cathode material (a metal) moving toward the positive pole, the anode (the cathode is here the negative electric pole, anode the positive pole). That motion of negative charges, which indeed seemed to be unidirectional, is thought to electrically polarize the gas molecules, resulting in the emission of light.

At this point, we naturally ask ourselves again whether this process can also be understood as a charge-separating, polarizing branching process. The voltage makes a charge separation in the gas molecules, creating opposite electric charges, whose *fusion* – the reversal of the electrical branching process – leads to the emission of light. If so, there should also exist an electric counter-current. That this is actually the case was demonstrated by GOLDSTEIN who detected positive charges moving towards the cathode, which he called 'channel rays' (he bored a hole, a channel, in the cathode material through which positive charges could escape). Interestingly, the lower the pressure of the gas, the better the gas proves to be polarizable. This seemed to strengthen MAXWELL's idea: If light can propagate in the universe, the so-called vacuum must contain an extremely rarefied gas.

Seventy years ago FARADAY had already experimentally determined the magnitude of an equalizing charge lacking for electrical neutrality as the so-called *elementary charge*. This was the smallest electrical symmetry disturbance that could be measured. However, FARADAY never called this elementary charge a particle, because the body idea of mechanics says almost nothing about the true nature of matter, which seemed to be explainable only by field ideas. Often the body is only a point abstraction without geometrical extension to which a mass is assigned. Ultimately, field theory was intended to provide a better understanding of the structure of light and matter and their electrical and magnetic properties. However, shortly after CROOKE's lecture, STONEY christened the smallest negative electric charge unit the "electron", implying a particle conception. In Germany, HELM-HOLTZ poetically called FARADAY's smallest electric charge an "atom of electricity" in 1881. In this way, atomistic associations crept in.

LENARD modified the glass cathode tube by cutting open the end of the glass bulb and sealing the opening with a thin aluminum foil. In this way, a negatively charged electric beam could be coupled out and used for experiments, just as polarized light beams are used in optical experiments. He found that cathode rays and light-induced currents were practically identical and absorbed by matter in the same way: The absorption depended only on the mass and thickness of the substance, not on their chemical or physical nature. LENARD concluded from this that electric rays interact with matter as if they consisted of electric charge units of certain kinetic energy, which he had called *electrical quanta* already in 1888. He was also able to prove that electric rays can be deflected by magnetic fields, which his teacher HERTZ had still denied. Therefore, electric rays had to be surrounded by a polarized magnetic field, just like a current-carrying wire. Since electric rays interacted with matter pulse-like and in discrete charge units (quanta of electric energy), the question arose whether they might not be better understood as longitudinal waves. But ten years later, in 1898, this concept was completely displaced by the idea that electric rays are electron-particle rays.

Light-induced electric currents and cathode rays could not be interpreted as light waves because they consisted of negative electricity and always emitted their energy in pulses in discrete amounts. After all, light should be electrically neutral and transmitted continuously, according to MAXWELL. This should also apply to the electrodynamic waves, which HERTZ had mistakenly called 'electric' waves, since they induce electric polarizations in conductors and dielectric polarizations in insulators (however, if the concept of bidirectional current is correct, they make dielectric polarizations in both cases). The electric induction effect was thus only the name-giver for these waves. But according to FARADAY, electric induction is induced by magnetic fields. Consequently, radio waves and visible light can just as well be interpreted as magnetic fields or waves. The term magneto-electric induction clearly expresses that the change in the intensity of the magnetic field at the location of the matter is the cause of the electric field that unfolds in or around the matter (which also appears as charge separation, dielectric polarization, voltage pulse, or electric current). Thus, while magnetic fields and light waves may well be the same thing, cathode and electron rays cannot be compared to light rays because of their charge and their pulse-like effects reminiscent of longitudinal waves. Initially, after the experiments of HERTZ, it seemed to be clarified that light propagates in the form of transversal waves and is continuous in nature, which should lead to an equally continuous transfer of light energy to matter. At that time, however, it was not yet known that the wave theory must be incomplete: only a few years later, the energy transfer of light also had to be described in terms of pulse-like effects – which led EIN-STEIN to remark that light behaved *as if* it consisted of particle-like quanta of light energy. Thus, the original idea of HUYGENS, YOUNG and FRESNEL that light consists of longitudinal pressure and pulse waves could have been rethought at this time. However, this would have required a thorough revision of optical wave theory and a new interpretation of polarization, which, as we now know, was the reason FRESNEL replaced the longitudinal wave model with the concept of transverse waves between 1821 and 1823.

Summary:

So, in reality, there is definitely a commonality between light rays, cathode rays, and gaseous matter rays that can possibly be interpreted with a puls-like acting longitudinal wave model. The branching process in quantum physics, which makes atomic notions and particle concepts definitely impossible, literally forces us to do so. So the crucial question is whether a modified longitudinal wave model can be reconciled with branching and polarization processes, which the experiments show so clearly. This is the same question that FRESNEL and YOUNG faced two hundred years ago. The simplest solution would have been to interpret the structure of matter and light as moving and standing longitudinal waves, which under certain circumstances can physically branch and form holistic field structures. As already mentioned, this branching process could well be called transversal, because it generates new spatial directions or 'degrees of freedom' which are always transverse to the direction of propagation (no matter in which direction a tree sprouts branches, these directions are always more or less transverse to the direction of growth of the trunk). So, in this picture, the branched partial rays could thus remain longitudinal waves that, when they interact or reunite, transmit pulse-like actions. The branches are coherent and form a physical-holistic structure. Since we know that a branched ray, wave or field structure is magnetically oppositely polarized in itself, thus embodies a mirror symmetry, we could also speak of a *holistically divided magnetic field*. Thus, in such a conception, a polarized light wave is nothing else than a magnetically mono-polarized partial ray of a branched, dipolar magnetic field. And a cathode ray is nothing else than a magnetically polarized partial ray of a branched electric field. The opposite polarized partial ray should then, as we know today, consist of *positrons* (the mirror form of electrons discovered in 1932), which in the case of the cathode ray tube are presumably dissipated via the grounding of the electric circuit. These are, of course, the missing positive charges in MAXWELL's electric current theory. In this way, branched light rays can be interpreted as magnetically polarized longitudinal waves which, like sound and pressure waves, come into effect in a pulse-like manner, but – due to the energy and symmetry preservation condition – not simultaneously, but only at one of the two ray tips. This in turn means that during the energy transfer, interactions must take place within the branched system, which can have any extension. We will return to this when we discuss Einstein's quantum hypothesis of light on the basis of partial reflection experiments.

The idea of tracing the propagation of light and sound to the same physical principles was already the goal of the first wave theory of light. This theory, in turn, was closely related to the theory of gases, as it tried to understand light as a manifestation of gaseous matter in motion. This approach was blocked by the phenomenon of polarization, which seemed incompatible with the hypothesis that light consists of longitudinal pressure waves in gases. As EINSTEIN once put it in the "Evolution of Physics":

"Well, it would be very welcome if we could see longitudinal oscillations in the light waves. Then the mechanical model of the aether could be constructed much easier. Most likely, our idea of the aether would then resemble in a certain way the mechanical scheme for gases with which we explain the propagation of sound waves."

This comment raises the question of whether we can fully understand the propagation of sound waves in gases within the usual mechanical scheme at all. The answer, of course, depends on how we interpret AVOGADRO's molecular division – mechanically or holistically. This directly affects the kinetic theory of gases and the theory of heat. Let us recall MAXWELL's attempt to understand light as the motion of a gaseous matter. He attributed the pressure fluctuations of the aether gas to electric and magnetic polarization processes, but could not understand their non-mechanical nature. We have found that it would have been quite possible to understand the transverse waves as holistic molecular division processes that magnetically (di) polarize. However, the idea that nature can perform holistic division processes did not even exist at that time. In physics, this idea still does not exist today, although quantum physics pretty clearly calls for it with the mathematical superposition postulate. So it is only a matter of taking the superposition postulate physically seriously. That means simply to recognize YOUNG's division condition unconditionally.

The original program of wave theory failed because the mechanical properties of the aether, i.e. the nature of the structural (polarization) movements, could never be elucidated. Now we find that this problem is closely related to the misinterpretation of molecular division and the reintroduction of the atomic hypothesis into molecular gas theory. EINSTEIN, who, despite his foresight, could never doubt the indivisibility hypothesis, concluded that the original program of physics (to explain Nature with the ideas of motion of mechanics) was not consistently feasible, and, as quantum physics proves, will never be convincingly feasible. What he meant by this, without really understanding it (like Bohr and Heisenberg), was the strange failure of the particle conception in the double-slit experiment, i.e. of the indivisibility or atomistic assumption, and the total failure of the equations of motion in the quantum mechanical interpretation of BOHR and HEISENBERG, which consequently could no longer be *Newtonian* mechanics. We notice again at this point that EINSTEIN means with 'mechanics' the classical mechanics of NEWTON, which does not know branching processes yet. If we introduce them, we get a new mechanics, i.e. a *new theory of motion*, which is primarily based on branching processes and dualistic principles of nature. This in no way devalues Newtonian mechanics, but adds to the old theory of motion (also called kinematics) a new form of motion that describes structural changes that would otherwise remain hidden forever. Only in this sense one can speak of a non-mechanical movement. Against this background, the enigmatic structural nature of radiation and matter can be discussed and evaluated anew.

The Michelson- Morley interference experiment

Towards the end of the 19th century, however, it became increasingly clear that something could not be right with the wave theory: Either with the aether concept, with FRESNEL's transverse waves, or with the nature of the motion of light. The first thing that became clear between 1881 and 1895 was that MAXWELL's definition of the speed of light led to inexplicable problems of understanding. This became clear by an interferometer experiment, which should prove the motion of the Earth through the aether gas.

An interferometer is an experimental setup in which a light beam branches into two coherent partial rays at a semi-transparent mirror or a simple glass pane placed at an angle of 45° to the incident light beam. The two partial rays propagate in different directions forming an angle of 90°: one ray is transmitted, the other is reflected. This is *partial reflection*, which clearly shows a bifurcation. In the interferometer setup, the partial reflection experiment is further supplemented by two full mirrors that reflect the two partial rays back into themselves. On their way back, the reflected partial rays hit the glass pane again, where they branch out further, superimpose with the original light rays and create interference stripes on a cardboard screen as in the double-slit experiment (Fig. 28, next page). So the two experiments must be closely related.

According to the theories of FRESNEL and MAXWELL, MICHELSON expected a *phase shift*, i.e. a displacement of the partial waves against each other, caused by the motion of the Earth through the aether. In the aether gas, the wave has the speed c, but with respect to the moving Earth, the speed of these waves should appear either slower (if the direction of motion is the same) or faster (if the direction of motion is opposite). Thus, if the Earth is moving in the same direction as the partial ray in interferometer arm A, the velocity of the Earth must be subtracted from the velocity of light to obtain the relative speed of light between the two systems, which would appear slower. This would have to be accompanied by a stretching of the wavelength in partial ray A, but not in partial ray B. The difference in wavelength should result in a shift of the interference stripe pattern. So much for the theory.

In 1881 the American MICHELSON carried out this experiment in Potsdam (Germany) and tried to improve it further with MORLEY, for the experiment gave an unexpected result: The expected shift of the interference stripe pattern did not show up despite sufficient resolution. This seemed to indicate that there was no speed difference between the split partial rays – or between the two branches of a holistically divided light beam. Neither the direction of motion of the Earth, nor the rotation of the interferometer caused any change of the interference pattern. So the speed of light had to be the same in both arms of the interferometer, regardless of its orientation and of the motion of the Earth relative to the aether. This raised profound questions: How was this compatible with the aether hypothesis? Was the wave model possibly wrong? Was it a standing wave? Hadn't one yet properly understood what happens when a beam of light divides?

To explain the null result of the experiment, some physicists suspected that the aether medium is dragged along at the Earth's surface, according to a theory already put forward by FRESNEL. Others suggested that the idea of a gas in the universe was a great fiction. But how then can the existence of *polarization waves* be explained, which HERTZ had clearly demonstrated? Was the real problem the concept of a unidirectional motion, which was the same in the theories of NEWTON and MAXWELL? Or does the Earth not move at all – neither through the aether nor through an empty Newtonian aquarium? Surprisingly, no one wanted to go back to the PTOLEMAIC world view, in which the Earth rests in the center of the universe, although this would have been a great opportunity – even the Catholic Church gracefully declined!

Starting point of all considerations was the fundamental assumption that waves in the aether medium always propagate *unidirectionally* with a constant velocity c, independent of wavelength and frequency. Therefore, moving light sources were expected to show the same frequency effects as sound waves, except that light should consist of transverse waves and visible light has wavelengths millions of times smaller and oscillation frequencies millions of times higher. Wavelength and frequency are in an inverse relation, because the velocity of light is just defined in this way: As the number of wavelengths – aka complete polarization cycles – passing a certain point per second. This model gives the impression that one could see, count or measure the passing polarization waves without disturbing them or absorbing their energy. However, as it turned out in quantum physics, this was a fallacy. Every interaction between light and matter must lead to structural changes on both sides.

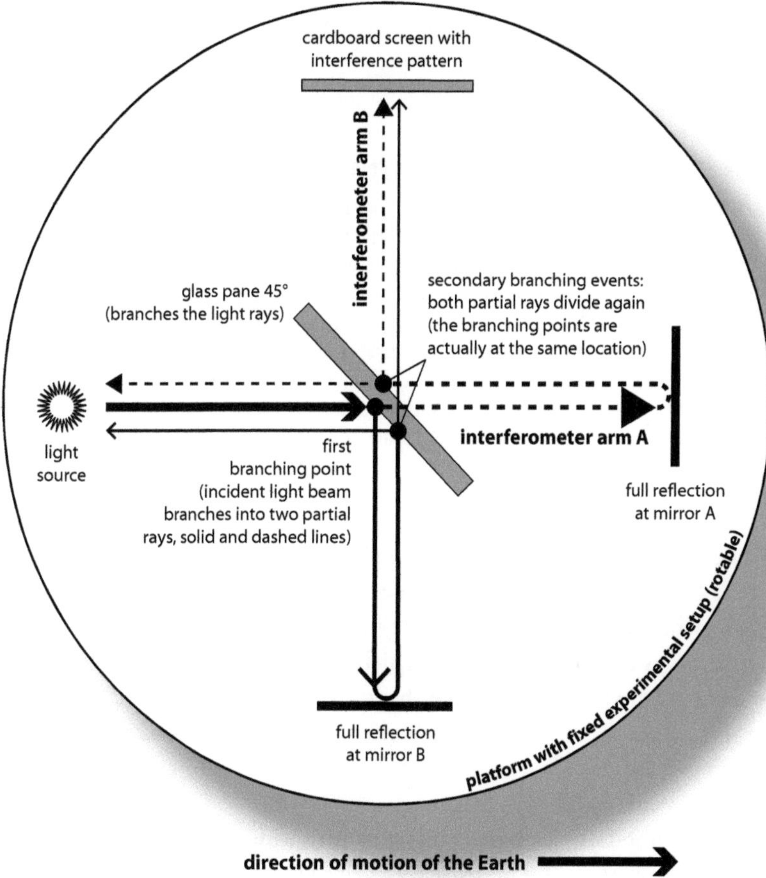

Figure 28 Michelson Morley experiment 1 (schematic diagram)

According to the theory of light, the wavelength of the partial rays in the interferometer arm A should increase if they propagate in the same direction in which the Earth is moving. The partial rays in interferometer arm B, however, should remain unchanged. Thus, the motion of the Earth through the aether gas should lead to a shift of the interference stripes. But neither the motion of the Earth nor the rotation of the interferometer platform caused a change of the interference pattern. It follows that there is no speed difference and change of wavelength between the two partial rays (solid and dashed).

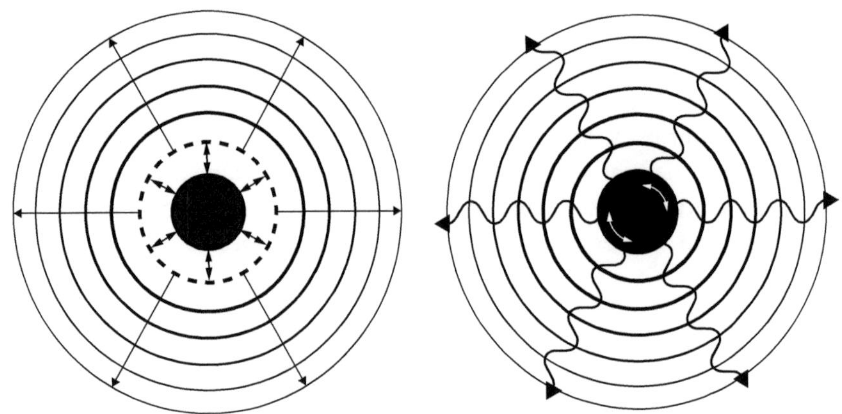

Figure 29 Spherical wave propagation: left longitudinal waves, right transverse waves

Left: A stationary, regularly pulsating ball (light or sound source) makes spherical longitudinal pressure waves in a fluid medium. Right: A stationary ball (light or sound source) rotating regularly back and forth generates vortex-like transverse waves in a liquid or gaseous medium like a whisk. The assumption that transverse waves also propagate as spherical waves was taken over from the longitudinal light and sound wave theory. This concept was naive on the one hand, since one had to assume now that the aether waves are generated by a back-and-forth rotation of the atoms in a viscous medium by friction. On the other hand, this idea was contradictory, because a ball rotating back and forth around an axis cannot produce spherical waves of uniform wavelength (they become smaller towards the poles). Moreover, a gas cannot be extremely dilute and at the same time highly viscous and, in addition, have an elasticity and stiffness that would have to far exceed that of steel.

The underlying difficulties can be better understood if one recalls that the theory of spherical wave propagation of light was adopted from the theory of sound propagation in ordinary gases (Fig. 29). It involves the Doppler effect, which is known from personal experience by anyone who has ever heard a passing fire engine: when it approaches, the siren tone gets higher, when it moves away, the siren tone gets lower. The theory of sound propagation in gases (like air) explains this as follows:

The vibrations of a membrane or pulsating body strike the surrounding elastic gas and cause collective vibrations of the molecules, which transfer these vibrations to the neighboring molecules in the form of pressure waves. These waves propagate through the elastic gas until the kinetic energy is consumed. They are longitudinal waves because the direction of vibration of the gas molecules is in the direction of propagation of the pressure waves. If we are dealing with a pulsating ball, the result is a longitudinal puls and pressure wave that expands spherically in the gas mixture we call air.

What exactly happens with the gas at the molecular level is a matter of the molecule theory and the kinetic model. In the particle theory of gases, sound and pressure waves are assumed to cause only elastic longitudinal vibrations of molecules and do not have the energy to change molecular structure or even split molecules. It follows that the motion of the sound source or the receiver leads to changes in the *effective* wavelengths in the elastic medium.

When the sound source is stationary, the waves propagate uniformly in all space directions in the form of spherical wave fronts (isotropic propagation). These are longitudinal pulse or pressure waves. Speed, wavelength and frequency of the pressure waves are the same in all directions, but they are gradually damped by the interaction with the elastic medium (Fig. 29, left). The same spherical propagation should also apply to transverse aether waves (Fig. 29, right).

When the sound source moves in the gaseous medium, the wavelengths in the gas change. In the direction of motion, they become shorter, while the waves radiated backwards become longer. This in turn changes the effective frequency, i.e. the time intervals in which the wave fronts arrive at stationary obstacles and receivers, although the wavelength and frequency of the source remain the same. Thus, in the direction of motion, the frequency increases at the location of the receiver. The tone pitch becomes higher because the puls waves arrive at shorter time intervals. This applies likewise to a wall, tree or other obstacle that is hit by the pressure waves. For pressure waves emitted in the backward direction, the distances between the wave fronts increase and the effective frequency decreases: the tone pitch becomes deeper for the receiver (Fig. 30).

The effect is the same when the source is stationary and the receiver is moving. Although the pressure waves from the stationary source propagate spherically with the same wavelength, these wavelengths arrive at the receiver at shorter time intervals as it moves toward the source: the effective wavelength becomes shorter, the tone frequency higher. As the receiver moves away from the source, these wavelengths arrive at dilated time intervals: the effective wavelength becomes longer, the tone frequency lower (Fig. 31).

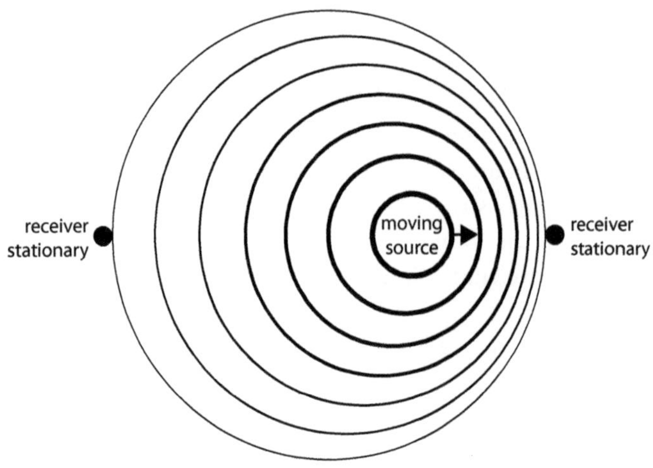

Figure 30 Doppler Effect: changing of wave periods due to motion of the source

When a regularly pulsating ball (light or sound source) moves in the medium, it moves with a certain speed relative to the speed of the spherical pressure waves emitted by itself. In the direction of motion, the waves are compressed (the wavelengths become shorter) and arrive at the location of a stationary receiver in shorter time intervals. The waves emitted in the opposite direction are stretched (the wavelengths become longer) and arrive at the location of a receiver in dilated time intervals.

This representation describes the propagation of sound waves and the frequency effects for receiver and obstacles as a function of the change in distance between transmitter and receiver in a gaseous medium at rest. The wavelength changes in the gas and the effective frequency changes for receivers are therefore objectively real and are a consequence of the relative motion between sorce and receiver.

This reminds of the old philosophical question whether a falling tree causes a sound when nobody is there to hear it. Of course, the falling tree makes a sound because it makes pressure waves in the air. So these pressure waves exist in reality, but in order to be able to perceive them, or interact with them, a membrane or material object is required that can be excited to vibrate, a receiver. In sound wave theory, the interaction between two objects can thus be explained by resonance phenomena.

The same physical principles should apply to moving light sources, even if the light should consist of transverse waves. In wave theory, the Doppler effect is symmetrical. It does not matter whether the source or the receiver or both are moved, the effect of changing

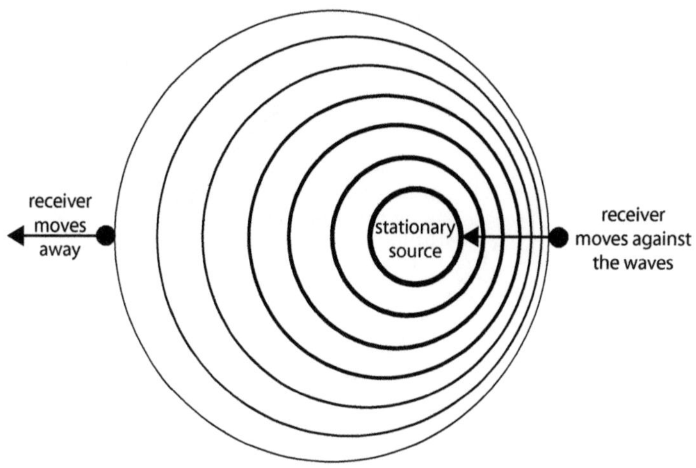

Figure 31 Doppler Effect: changing of effective wave periods due to motion of the receiver

This is what the stationary spherical wave looks like for a moving receiver: If it moves toward the source, that is, against the propagation speed of the spherical waves, the waves will strike the receiver in shorter time intervals. The wavelengths appear shorter (although the wavelengths emitted by a stationary source do not change). If the receiver moves away, the waves arrive at dilated time intervals, the wavelengths appear longer. Thus, we are dealing with effects that require a change in distance.

the effective wave period – the time interval of the impinging waves – depends only of the change of the distance. This reminds us of the problem to explain FARADAY's induction symmetrically, which even MAXWELL did not succeed in: the induction process must be the same (invariant) whether one moves the conductor or the magnet; therefore the physical model and the corresponding equation must be able to represent a symmetry of the motion. This is basically a bidirectional approach as well. And it is exactly at this point that all the confusing complications arise, which should eventually lead to the special theory of relativity.

Could physical branching processes also play a crucial role in this experiment? Obviously, this was an experimental situation in which the notion of a unidirectional motion of light was to be applied to a *branched* ray system. The light source, the branched light beam and the interferometer form a system that is pivoted on a plate on the Earth's surface and moves with the Earth either through an empty universe or an extremely rarefied gas. The light beam branches at

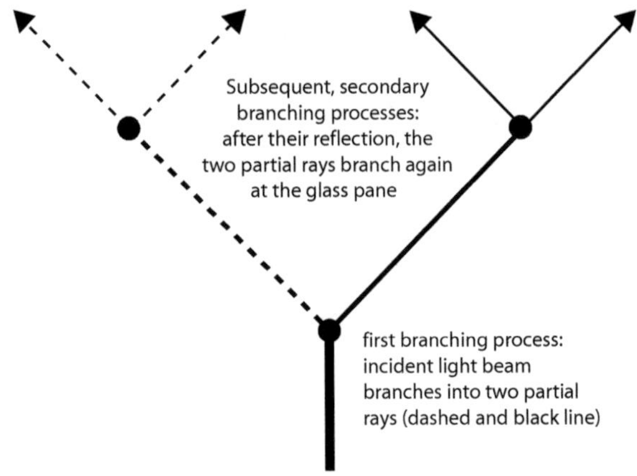

Figure 32 Michelson-Morley experiment 2 – branching structure

If there are physical branching processes in reality, the following picture emerges: the incident light beam branches at the glass pane (first branching point) into two partial rays, dashed and solid. At the secondary branching points, the two primary partial rays each split again into two secondary partial rays. Two of them interfere and form the stripe pattern, the other two are reflected back to the light source (not shown here).

the beam splitter (due to birefringence at the glass pane) and the two partial rays propagate *independently* from there again at the speed of light, at least according to the theories of FRESNEL and MAXWELL. The mirrors serve only to extend the light paths and to map the two subsystems onto each other so that they can superimpose and produce an interference pattern. If the presuppositions are correct that light *propagates* (1) in the form of *transverse waves* 2) in a staionary (3), elastic, continuous *gaseous medium* (4) with *constant velocity* (5), a relative motion between the source and receiver must change the effective wave period at the receiver. This is the analogy to the acoustic Doppler effect. However, the optical Doppler effect likewise applies only to a relative motion between transmitter and receiver, i.e. a *change in distance*. So the Doppler effect is also applicable to light, but apparently not in this case. Why not? This was the great question.

Is this because there is no relative motion between the transmitter and receiver in the MICHELSON experiment, since both are on the same platform? The only relative motion that can be assumed here is that between the moving Earth and the immobile aether gas in which the light waves are supposed to propagate (independently of the light source which made the waves). Is this not the same case as in Fig. 31, where the receiver moves relative to the waves excited in the gaseous medium by a source just before? If so, why is there no change in the effective wave period in the direction of motion?

Fact is that the MICHELSON-MORLEY experiment could not prove any changes of wave periods. If all five assumptions are correct, i.e. the theory of light is correct and complete – is it conceivable that these changes of the wave periods actually take place in reality, but are compensated by a yet unknown physical process? Is the situation even comparable? After all, in this experiment we are not dealing with a transmitter and a receiver moving relative to each other. Moreover, there is no receiver at all, only an interference stripe pattern on a detection screen, which could also be a simple piece of cardboard. Can this be considered a receiver?

Obviously, we are dealing here with an interference experiment in which a light beam is branched at a glass pane, each partial wave is reflected back on a full mirror and sent back to the beam splitter, where it is branched again. The whole system is therefore a multiple branched field mirrored in itself. Is this possibly a *standing wave* that does not move at all with respect to the light source and the beam splitter? This is exactly the same situation as on the laboratory table. Consequently, the motion can have no influence on this branching structure. So, if we imagine that this experimental arrangement is moved through the universe and we want to make any statements about it, we have to interact with this system physically – at least by the exchange of light rays. But this must change the structure of the light, as we know from the quantum theory. In the second volume we will see that we need to understand partial reflection in quantum physics, i.e., the branching of a single ray of light carrying a certain amount of energy, and why a branched ray can transfere its energy to matter at only one of the two possible interaction points. Then we may also understand what happens in a deeper branched system.

However, the problem of determining the relative speed between two moving reference systems interacting via the exchange of light rays remains even if one rejects the aether gas hypothesis. If light is assumed to propagate at a constant speed in a vacuum, NEWTON's addition theorem of velocities must still hold: the local speed of light must still depend on the direction and velocity of the moving matter – at least if this matter interacts with the light. This shows – as EINSTEIN was the first to recognize – that the principle of symmetry of motion, which plays such a fundamental role for the description of force-free relative motion and the definition of inertia in Newtonian mechanics, is still missing in the theory of light.

The Fingerprint of Matter

Since FRAUNHOFER (1814), KIRCHHOFF (1859) and BUNSEN (1859), visible light has been studied by means of *spectral analysis*. Initially, glass prisms were used, later increasingly finer diffraction gratings. Newton had discovered that a glass prism splits the 'white' sunlight into rays of different colors, but it was FRAUNHOFER who discovered that this continuous band of color also contains dark lines that only become visible on closer inspection and better resolution. He found 567 dark lines in the spectrum of sunlight and designated the most conspicuous of them with capital letters. One particularly conspicuous double line he called the D-line. It turned out that the black lines are caused by the absorption of certain wavelengths of light by gases. In other words, light rays of a certain color are absorbed by matter, which indicates structural changes of the molecules. Since YOUNG and FRESNEL were able to assign specific wavelengths to the color spectrum based on the spacing of the slits and the strength of diffraction, it was known that the wavelength in the visible light spectrum decreases continuously from red to violet, while the frequency and therewith the energy content of these waves increases in inverse proportion. So red light has longer wavelengths and lower frequency and energy, while violet light has shorter wavelengths and higher frequency and energy. And since HERTZ is was known that at the red end, the spectrum continues with invisible infrared light, microwaves, and radio waves, reach longer and longer wavelengths, which contain correspondingly less energy. At the violet end of the spectrum follow the invisible, high energy, extremely short wavelength ultraviolet, gamma, and X-rays. And due to refraction and diffraction phenomena, it was possible to accurately determine the wavelengths and frequencies of light rays.

When KIRCHHOFF and BUNSEN analyzed the light from the flame of a Bunsen burner in which they vaporized sodium (common salt), they discovered two bright yellow lines exactly at the position where the dark double D-line appears in the sun spectrum. Further investigations confirmed the idea that each chemical element emits light at typical frequencies and wavelengths and produces a characteristic line spectrum, i.e. no continuous spectrum. However, they also found that a glowing, heat-radiating body emits a continuous spectrum of light, just like the sun. The spectrum of thermal radiation was the same for all glowing bodies, so strangely enough it did not depend on the substance it contained. This insight inspired them to send the light from a glowing body through the sodium-enriched flame. They found that the two yellow D-lines now appeared black, as if they were cut out of the continuous spectrum of the glowing body, just as in the Fraunhofer spectrum of sunlight. From this they concluded that heated sodium vapor absorbs exactly the light rays that it would otherwise emit as a yellow double-D line. The sodium molecules (whether atoms really existed was not yet clear) could therefore either emit or absorb these wavelengths, but not both at the same time. So in the first case they had produced an *emission spectrum* of sodium (bright lines), in the second case an *absorption spectrum* of sodium (dark lines in continuous color spectrum). The lines are located in exactly the same places. Thus, the emission and absorption of light by molecules or atoms always occurs at the same wavelengths and frequencies that are characteristic of the chemical element in question. This discovery enabled KIRCHHOFF and BUNSEN to identify both the yellow and the dark double line with the presence of the element *sodium,* in the first case in the light source, in the second case in the gas or vapor between the light source and the spectral analyzer. Further research confirmed that any chemical element, whether it is a glowing gas or matter vaporized to a gas-like state, emits light in the form of a typical line pattern that is discrete, i.e., non-continuous in nature. Thus, it makes no difference in principle whether the gaseous state consists of double molecules or integral molecules (our atoms). More complexer molecules, however, produce so-called band spectra in which sharp intensity lines can no longer be identified.

During a total solar eclipse in India in 1868, the French astronomer JANSEN succeeded in proving that the solar corona emits a bright sodium double line. Consequently, the hot gases of the corona, the outer solar envelope, must contain, among other elements, the element sodium. Normally, however, without a solar eclipse, one registers the dark Fraunhofer D-lines in 'direct' light, i.e. in the continuous solar spectrum. From this developed the theory that the sun must have a glowing core that emits a continuous thermal spectrum, and a glowing gas envelope that absorbs certain light rays from this continuous spectrum, producing the dark absorption lines. Consequently, by analyzing emission and absorption spectra, it was possible to infer the chemical elements in the radiation source and in the gases through which the light passed. The resolution of spectral analyzers improved rapidly. Physicists examined all known chemical elements and could soon detect even smallest amounts of chemical elements. In this way, some new chemical elements were also found: Rubidium and Cesium by BUNSEN and KIRCHHOFF (1861), Helium in the sun by JANSEN (1868), Gallium by BOISBAUDRAN (1875), and Helium on the Earth by W. RAMSEY (1895). It was even possible to study the light of stars with telescopes and spectral analyzers, so that statements could be made about the presence of chemical elements in distant suns. It was unbelievable: Light provided a characteristic fingerprint of matter that could be used to accurately determine the chemical elements in light sources and glowing gases – whether on the lab bench or billions of kilometers away!

So the wavelength or frequency of the emitted and absorbed light rays had to be directly related to changes in the molecular structure of matter, which also had to be electromagnetic in nature. And if the hypothesis was correct that molecules were composed of atoms, then the hypothetical atoms, which until then had remained only vague billiard ball abstractions of the thinking mind, must themselves have an electromagnetic structure. This remains true even if atoms are nothing else than integral molecules that are holistically divisible, as in our interpretation of AVOGADRO's molecule division hypothesis. One could certainly try to understand this with a purely electromagnetic field theory, in which fields can divide holistically. In the simplest picture, these are single light rays which branch.

But this idea did not exist yet. Optics could not explain why light beams or waves branched at all. There was no dynamic theory that could explain this. This is exactly what MAXWELL failed at when he tried to explain the nature of the aether with elastic-plastic deformations by forces he could identify as electric and magnetic polarizations. So one had to be content further with kinematic, i.e. force-free descriptions of the behavior of light beams. Since polarization was interpreted as unidirectional motion in the sense of Newtonian mechanics, the idea that polarization might be a branching process representing a new type of kinematics was not exactly obvious. This in turn steered the emerging electromagnetic theory of molecules and still hypothetical atoms in a specific direction: one now tried to combine the electromagnetic phenomena with the atom idea and mechanics. Paradoxically, to do this, one first had to postulate new, even smaller, again indivisible particles in the supposedly indivisible atom. This process began in 1896 and was to become a real national sport in 20th century physics. However, an ultimate 'atom', a truly indivisible particle, could never be found.

Faraday rotation: The magnetic branching of polarized light rays

In 1896, when ZEEMAN was investigating certain magneto-optical phenomena, he remembered an old experiment by FARADAY. FARADAY had the idea of exposing a sodium flame to a magnetic field in order to find out whether this caused the two light beams that produce the double D-line to branch out further. This was the last experiment FARADAY had performed in 1862 without being able to demonstrate the suspected magnetic induced splitting of the spectral lines of the emission spectrum. With stronger electromagnets, however, one should be able to observe the doubling of the spectral lines, with the magnetic field splitting the light beam into two partial rays with opposite circular polarization. This was at least the explanation KELVIN (WILLIAM THOMSON) had proposed to explain the FARADAY effect or FARADAY ROTATION, also known as *magneto-optical Faraday effect*. This branching was mathematically interpreted as splitting into two partial rays with opposite properties in the sense of vector decomposition (already by STOKES), which was later to be called the *superposition principle*. However, a real branching process or new physical principle of nature was not recognized in it, since the transverse wave model and the composite-hypothesis seemed to explain the phenomenon of polarization sufficiently well. At this point, physics has overlooked its greatest discovery for the second time (the first time it was the true nature of the molecular division process).

In the FARADAY rotation experiment, a ray of light, already linearly polarized by partial reflection, passes through a square lead glass pane with polished edges (approx. 5 x 5 cm, thickness 1.3 cm). The ray enters at one edge and exits at the opposite edge. The longer the distance of the glass to be traversed, the more pronounced the effect. Thus, the mass of the dielectric matter must play a role in this magnetization effect. The outgoing ray hits an eyepiece equipped with a rotatable nicol prism that analyzes the polarization of the outgoing ray. The opposite poles of a strong electromagnet are placed on the opposite edges as close as possible to the light ray so that the magnetic field lines are parallel to the light ray and the glass pane (later, magnetic poles with holes were used so that the ray can pass exactly parallel to the field lines). If the electromagnet is switched on, the plane of polarization of the outgoing light ray rotates. If the polarity of the magnet is reversed, the plane of polarization rotates in the opposite direction. The amount of rotation depends on the strength of the magnetic field and the distance traveled (Fig. 33).

Next to the double-slit experiment, this is the most important key experiment in physics. However, the assumption that it shows the rotation of a 'wave plane' is just a theory-driven conclusion. It is not that you see a line of light that changes its angle or something like that. The only thing that rotates here is the nicol polfilter (a double

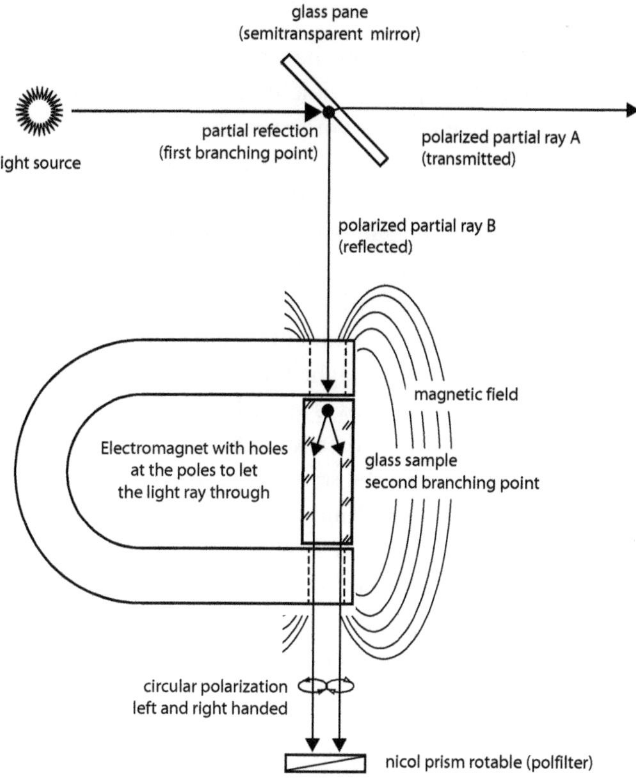

Figure 33 Faraday rotation: The magnetic branching of light creates an enantiomorphic field

The Faraday rotation experiment provided the first evidence that a polarized light ray is magnetic in nature. It shows that a strong magnetic field can further branch an already polarized partial ray of sunlight radiating through a glass body. This results in a branched ray system with enantiomorphic properties consisting of two coherent partial rays. The mirror symmetry manifests itself in the complementary colors of the two partial rays and their left- and right-handed circular magnetic polarization. The superposition of both partial rays into one outgoing ray causes a rotation of the original polarization plane.

prism made of glued calcite crystal pieces), which acts as a polarization filter and analyzer and is attached to the eyepiece: it has to be turned to the right or left until the light shines completely through again (a polarizer works like a light dimmer if the light is polarized, i.e. you only see a more or less bright or even opaque nicol prism). So, when the electromagnet is switched on, the original translucent nicol polfilter darkens, but it can be rotated until the light shines through again. The acting magnetic field must therefore have created a perturbation in the glass material, an anisotropy, a new physical degree of freedom that changes the structure of the light beam and causes a rotation of the polarization 'direction' of the outgoing light ray. This new degree of freedom is *chirality*. This term refers to a holistic physical system with a mirror-symmetric structure, which can only be generated by holistic division and branching processes. At this point, this (still) sounds like an assertion, but it is the conclusion that the big picture of quantum physics forces us to reach. This apparently applies not only to humans (think of your right and left hand), but also to the structure of light: this process divides and doubles an already polarized light ray by creating left-handed and right-handed partial rays. This is what I called an enantiomorphic field. That it really exists is therefore also proved by the FARADAY rotation, which represents the starting point of the electromagnetic theory. In other words: behind the electromagnetic theory hides a still undiscovered, completely new physical principle. Obviously, behind the Faraday rotation is hidden a mirror-symmetric polarization of the aether gas, or, if the aether does not exist, of the 'free' electromagnetic field. So the branching process polarizes in the truest sense of the word: it causes a dimagnetic polarization of the incident polarized light ray, which divides into left- and right-handed partial rays with opposite magnetic polarization (which we model with opoosite magnetic field vectors).

The special lead-borate glass – a dielectric material – obviously plays an important role in this process: recall the dielectric polarization of a glass block, when a voltage is applied (p. 135). Now it is a magnetic field that appears to produce a dimagnetic polarization of the molecular structure of the entire glass block (which is a chirality or dipolarity that obviously structures the substance *holistically*), which in turn must be related to the fact that the incident polarized light ray branches out further. The piece of glass thus appears to act as a resonant and amplifying medium for the external magnetic field, which greatly enhances the extent of the magnetic dipolarization of the branching light ray. If this is true, the FARADAY effect should also be detectable on individual polarized light rays without dielectric or dimagnetic materials, but with much stronger magnetic fields.

Faraday observed the following: when the light ray shines through the glass sample and the eyepiece with the polfilter is rotated until the transmitted light shows its greatest brightness, the polfilter has a definite reference position. This is the zero degree position from which the rotation of the polarization plane is defined. If the electro-magnetic field is turned on, the polfilter darkens. If he turned it clockwise from this position, red light becomes visible; if he turned it counterclockwise by the same amount, greenish-blue light – the complementary color – becomes visible. Since red light has a longer wavelength (less energy) and greenish-blue light a shorter wavelength (more energy), Faraday concluded that the magnetic field must have split the incoming light ray into two partial rays with different frequencies and opposite polarization states. The oppositeness of the two polarization states was interpreted as right-handed and left-handed circular wave motion of the partial rays, since the polarization filter had to be rotated left or right from the determined zero position to determine their respective polarization directions. The position in the middle between the two states can be understood as the symmetry axis of the chirality of the two partial rays (in this position no light is transmitted, when the electromagnet is swiched on). At this time it was already known that circular polarization had to be interpreted as a summation of many individual, local polarization processes, each of which produced opposite polarizations in the glass substance. This meant basically that we have to do with holistic refraction and diffraction processes in which the light beam divides 'globally' into two oppositely constituted partial rays, which in turn experience the same processes locally: they are refracted and branched manifold at atoms or molecules, but the two partial rays always remain a whole. Since the resulting partial waves had to be of opposite nature, they were interpreted as transverse wave planes rotating in opposite directions during their propagation, thus forming helical surfaces.

Faraday's experiment made it now clear that a *polarized* light ray is susceptible to magnetic fields, so light had to have magnetic properties themselves. In his laboratory diary, Faraday noted in 1845 that he had finally succeeded in *"magnetizing a ray of light"*. In other words, the magnetic properties of light are caused by branching processes that magnetically (di)polarize. In the experiment, an already magnetically polarized ray of light is caused by the external magnetic field to branch again. Since the incident ray was already polarized, it could itself have been only a branch, i.e. a partial ray, of an already branched light beam. Pictorially, this can be summarized like this: we see here a branch of an already branched magnetic field, which branches again mirror-symmetrically in the glass matter under the influence of an extern magnetic field. That the branched ray system must be of holistic nature is shown by the fact that they interfere and act together energetically.

Mediator of the branching effect is the transparent dimagnetic or 'diamagnetic' glass material (as it was later called by FARADAY), acted upon by a strong magnetic field (polarized magnetic force) that creates a non-local, holistically organized magnetic dipole structure in the glass. Besides the double-slit experiment, this is the strongest indication for a holistic division: if it were only a mechanical division, a doubling with identical partial rays would have to result. But the FARADAY rotation experiment already shows that the branched rays are of mirror-symmetrical *magnetic* nature. However, this was not yet clear at that time. The task of the aether wave theory – and any other theory of light – was to explain why the magnetic branching process leads to a rotation of the symmetry axis of the chiral ray system and how the frequency differences of the partial rays come about.

If we now remember the shielding effect, it becomes clear that the magnetic field in the glass material must be of opposite polarity, i.e. representing a magnetic counter-field (p.133). We can see from this fact that *di*-magnetic and *di*-electric polarization processes in glass are very similar and possibly even identical, which was already suspected by FARADAY. This antifield is either the cause or the consequence of the branching process. However, it was not clear how the magnetic anti-field acts. As FARADAY found, when his glass sample was suspended between the magnetic poles so that it could rotate freely, it aligned itself transversely rather than longitudinally to the magnetic force lines. Apparently there was a magnetic shear force or torque that aligned the magnetic antifield in the di(a)magnetic glass body transversely to the extern magnetic field, while ferromagnetic substances like iron files always align parallel to the magnetic force lines. This was reminiscent of the transverse nature of the circular magnetic lines of force around current- carrying conductors, which have no magnetic poles but exhibit magnetic polarity. These strange properties led to a discourse between FARADAY and WEBER. WEBER argued that magnets cannot really be understood as dipoles. They merely reflected polar electrodynamic forces à la AMPERE, to which we can assign opposite magnetic poles only because of their *apparent* preferred direction. FARADAY, however, considered the existence of opposite magnetic poles and magnetic lines of force as proven and spoke of real existing magnetic fields for the first time at the end of 1845, twenty-five years after the discoveries of OERSTED and AMPERE. But the remaining inconsistencies, which could no longer be explained by mere attractive and repulsive forces, soon required the idea that magnetic fields influence space 'globally' and that the *effective* magnetic forces are 'local' in nature. At the same time, FARADAY introduced the terms paramagnetism and di(a)magnetism for the two forms of *induced* magnetic fields in non-conducting matter. A magnetic field can induce symmetric or antisymmetric magnetic fields in dielectrics (para: same force direction, anti: opposite force direction). Thirty years later, 1875, Prof. TAIT summarized the theory behind FARADAY rotation as follows:

"The explanation of Faraday's rotation of the plane of polarization of light by a transparent diamagnetic requires, as shown by Thomson, molecular rotation of the luminiferous medium. The plane polarized ray is broken up, while in the medium, into its circularly polarized components, one of which rotates with the ether so as to have its period accelerated, the other against it in a retarded period. Now, suppose the medium to absorb one definite wave-length only, then — if the absorption is not interfered with by the magnetic action — the portion absorbed in one ray will be of a shorter, in the other of a longer period than if there had been no magnetic force; and thus, what was originally a single dark absorption line might become a double line, the components being less dark than the single one". [31]

31 Quoted from Zeeman in the appendix: P. Zeeman: "On the Influence of Magnetism on the Nature of the Light Emitted by a Substance", March 1897.

ZEEMAN's comment: *"Hence here the idea is perfectly clearly expressed of the experiment"* (which FARADAY tried in vain to prove). Thus, the FARADAY rotation should be based on an enantiomorphic branching of the polarized light ray, resulting in the splitting and doubling of the spectral line. In TAIT's description you can also clearly see that the "broken up" light ray describes nothing else than a branching process. So there must be a connection between FARADAY rotation, doubling of spectral lines, enantiomorphic branching and polarization processes.

Recall now that the FARADAY effect was the starting point of MAX-WELL's electromagnetic aether theory. Assuming the validity of the transverse wave model one 'sees' a rotation of the wave plane which KELVIN interpreted as elastic (fully reversible) torsion of the aether substance in 1856, as a *"molecular rotation of the luminiferous medium"*. From FARADAY's experiment and FRESNEL's theory he concluded that the two helical waves must propagate with different speeds in the glass because of the frequency (color) difference, which can only be explained with a change of the wave periods of the partial rays. This change should be caused (as in the Doppler effect) by a relative motion, now a rotation, against the aether. KELVIN thus imagines a relative motion in which the branched ray system as a whole – a fork, so to speak, with two prongs rotating in opposite directions – turns in the resting aether. So KELVIN imagines a relative motion, where the branched ray system as a whole – so to say a fork with two prongs rotating in opposite directions – turns in the resting ether. Kelvin already noted that this was basically a nonlinear process (as we would say today) that must take into account dynamic feedback effects with the external magnetic field both within the aether and with the glass matter.

At this point one can ask oneself again, which picture arises, if one interprets AVOGADRO's molecular division process holistically and presupposes that the aether gas really exists. We then see a "molecular" process of division and branching of the gas, which generates an enantiomorphic magnetic field that responds to the external polar magnetic field by rotating about its axis of chirality. The word molecular appears here in quotation marks because we have so far imagined it to mean something extraordinarily small. However, as we have already suspected in the Avogadro chapter and can also see in this experiment, molecular division has nothing at all to do with size scales. It is a qualitative process which is the same with light and matter. If we nevertheless imagine the usual molecule picture, we see a molecular cell division which aligns the gas molecule in the external polar magnetic field due to its emerging chiral character. The double molecule now behaves like a magnetic dipole. From my point of view, this shows that the magnetic dipole character is only an expression of a branched magnetic field (in this case, of light), and that one must look at the branched system holistically to be able to recognize a chiral state in it. Therefore it is important to mention that in the FARADAY rotation experiment an unpolarized sunbeam branches into two oppositely polarized partial rays by partial reflection at an ordinary glass pane. One of these polarized partial rays is sent through the lead-borate glass where, under the influence of an external magnetic field, it branches again into two finer, oppositely polarized twigs. As can then be seen, chirality exists at each of these levels of consideration, down to molecular scales that describe inter-actions of the magnetic field with the glass material. This also shows that FRESNEL's composite-hypothesis of light beams, which had to presuppose definitive wave planes in order to explain polarization, can be easily replaced by the branching hypothesis.

If these considerations are not completely wrong, it becomes also clear why MAXWELL could not model the dynamic deformations of the aether gas and the phenomenon of polarization with purely me-chanical (i.e. unidirectional) motions alone. Thus, it would be quite worthwhile to try an interpretation of the FARADAY effect based on physical branching processes of light and gasous matter. At FARA-DAY's time, however, the splitting of individual spectral lines was not yet detectable. If this had been successful, it would have been the proof of MAXWELL's assumption that the aether gas – and a single ray of light – is dimagnetically polarizable like ordinary gases. It was to take another fitfy years before PETER ZEEMAN succeeded in 1896 in demonstrating the splitting of spectral lines in gases predicted by FARADAY.

The Zeeman Effect: The magnetic branching of matter

ZEEMAN followed FARADAY's reasoning and assumed that the magnetic field acts mainly on the molecules of the medium being traversed. Consequently, the splitting of a spectral line could be much easier to detect in the absorption spectrum. Inspired by the experiments of KIRCHHOFF, BUNSEN and FARADAY, he designed a special experimental setup to analyze the passage of a light beam through gaseous substances exposed to strong electro-magnetic fields. As a gas container, he used a porcelain tube that he could fill with various gases and chemical elements to be vaporized. Equipped with much stronger electromagnets and finest ROWLAND diffraction gratings, he was now able to demonstrate the splitting of spectral lines for a whole range of gases and vaporized substances. He found that the magnetic field splits the original absorption lines in a mirror-symmetrical way not only in double lines, so-called doublets, but also in triplets. Later, even quadruplets, quintuplets and sextuplets were discovered. Thus, the original spectral line splits into two or more spectral lines under the influence of the magnetic field. Since this occurs in the absorption spectrum, the cause must be a magnetically induced change in the molecular structure of the gas, which absorbs the incident light beam which has splitted into two or more polarized partial rays. The split spectral lines appearing to the right and left of the original line show that the magnetically altered molecular structure now absorbs the incident light beam at two, three or more different wavelengths or frequencies. Since absorption and emission spectra are identical, the process of the magnetic induced molecular structure formation must be reversible.

ZEEMAN enclosed the gases and chemical elements to be vaporized in the porcelain tube sealed with glass plates at both ends to allow a beam of light to pass through. The tube was about the size of a modern laser pointer and was heated with a Bunsen burner in a uniform rotating manner to avoid density differences in the gas. The heated gas then emitted light with a continous thermal spectrum. Then a beam of light was then sent trough the porcelain tube, interacting with the gas, which absorbs a specific wavelength (or two in the case of sodium). The outgoing light is focused into a beam with a lens, passes through an aperture with a vertical slit, and then hits

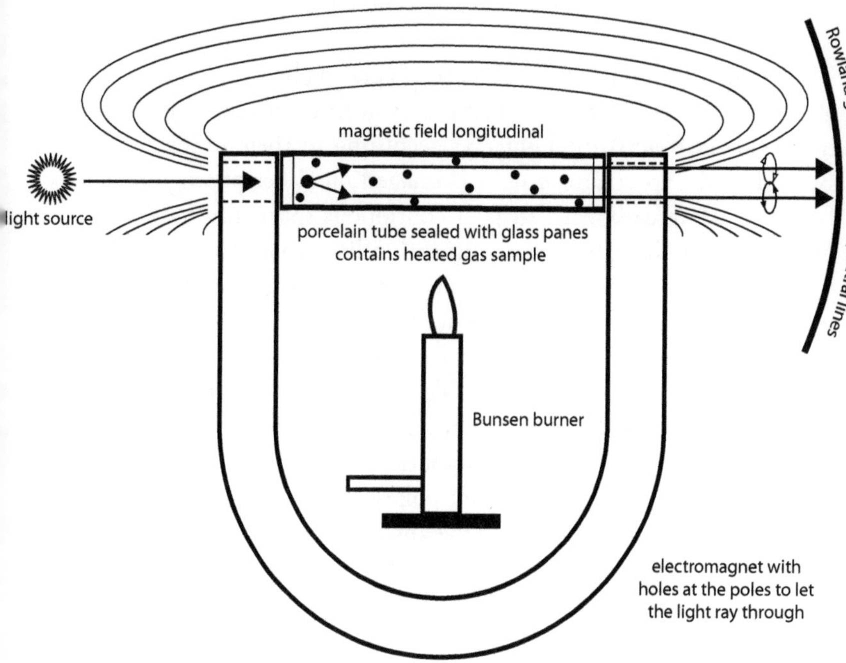

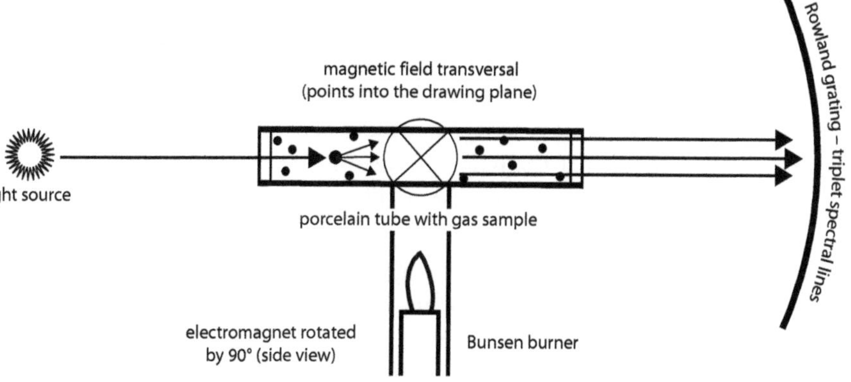

Figure 34 The Zeeman effect: the magnetically induced branching of matter and light

These experiments show that the molecular structure of gases can be branched by magnetic induction (as can light rays), as evidenced by the single or multiple doubling of absorption spectral lines. If the magnetic field is arranged in the longitudinal direction of the tube, the passing light beam branches into an even number of spectral lines (partial rays) as in the Faraday rotation experiment. If the magnetic field is arranged transverse to the tube, the passing light beam branches into an odd number of spectral lines (partial rays). The branching creates a new, mirror-symmetric ray structure: the split spectral lines are equidistant from the original line and deviate from the original frequency by (almost) the same amount.

a Rowland diffraction grating. ZEEMAN's grating had a resolution of about 15,000 lines per inch and a radius of three meters. The grating bends (diffracts) the light rays according to their wavelength. Behind it, the spectrum of light is imaged on either photographic film or a white cardboard strip. A continuous spectrum is shown, interspersed with vertical dark absorption lines that can be clearly distinguished with an eyepiece (note that the grating and the projection screen are mounted horizontally here). Their polarization is again determined with a rotatable nicol polfilter attached to the eyepiece. The fact that ZEEMAN was able to determine the polarization of *absorbed* light rays is initially puzzling, since the rays are extinguished. But the extremely high resolution of the diffraction grating made it possible to determine their polarization at the edges of the dark lines. Since the position of the absorption lines is wavelength dependent, they indicate at which wavelengths the light is absorbed by the gaseous medium in question (Fig. 34).

Strangely enough, the number of spectral lines generated depends on whether the magnetic field acts on the luminous gas in the longitudinal or transverse direction: In the first case, the magnetic poles are located at the ends of the tube (hence the pierced magnetic poles to allow the light beam to pass through). So in the longitudinal case the magnetic field lines have the same direction as the tube and the beam of light, they are parallel as in the FARADAY rotation experiment. There are two versions of parallel magnetic polarization, since the magnetic field can point in opposite directions, depending on how the magnetic poles are arranged (if polarization represents a space direction, which we doubt since MAXWELL's provisional model of a unidirectional mechanical motion). Since it is an electromagnet, the poles can be swapped at the touch of a button. In the second case, the poles of the electromagnet are arranged transversely to the axis of the tube so that the magnetic field lines cross the tube and the transmitted light beam at right angles.

That the magnetically induced change of the molecular structure, and not heating, is responsible for the splitting of the spectral lines is shown by the following fact: when the magnetic field is switched off, the doubled absorption line disappears and the original single line reappears. In the case of sodium, the quadruplet disappears and the original double D-line is restored. The external magnetic field must therefore change either the structure of the atoms and molecules or the molecular structure of the gas as a whole. This process is obviously reversible: if the magnetic field is switched off, the gas returns to its original (thermally excited) state.

In the longitudinal case (i.e. magnetic field lines parallel to the tube and the transmitted light beam), the original absorption line always splits mirror- symmetrically into an *even number* of spectral lines. They appear to the right and left of the original spectral line, which disappears, clearly indicating a branching of the light ray. This also shows that the wavelengths and frequencies of the absorbed partial rays differ by the same amount because the distance to the original center line is the same for both lines. ZEEMAN could also prove that the two lines are oppositely circularly polarized, one left- handed, the other right- handed. This corresponds to the branched rays in the FARADAY's rotation experiment, except that we are now speaking about absorbed light rays. If we consider them as 'shadow rays', they branch exactly like light. This in turn points to molecules that can branch (or divide holistically) like light. If the electromagnet is switched off, they merge back to the original absorption line with linear polarization.

In the transversal case (magnetic field lines across the tube and the transmitted light beam), the original spectral line always splits into an *odd number* of spectral lines. They are all *linear polarized*, but in a opposite way. The spectral line splits into triplets or quintuplets because the original absorption line does not disappear. The lines to the left and right are linear polarized, but orthogonally relative to the spectral line in the center (note that the spectral lines are mapped vertically here). So if the center line is horizontally polarized, the outer lines are vertically polarized, or vice versa. (I would even suspect that the outer lines are antiparallel polarized, but this is not provable with polarizing filters, since they cannot distinguish opposite polarization states in the same axis of chirality). The spectral line of the vaporized chemical element cadmium even splits into five spectral lines: a center line and two offset pairs to the left and right of it.

The symmetric spacing of the new emerging spectral lines shows that the wavelengths or frequencies of the splitted absorption lines are *energetically equidistant* from the original frequency. This applies to both longitudinal and transverse magnetic fields. The difference in frequency is therefore almost the same, *it differs only by its sign*. What one absorbed partial ray has more in frequency, the other has correspondingly less. And this is possibly a point where one can also think different.

According to MAXWELL, the nature of electric polarization (and of electric current) is a unidirectional displacement of exclusively negative charges in two possible directions. Hence, they are mechanical movements which take place one after the other. The reciprocating motion of the negative charges induces transverse aether waves with analogous oscillating magnetic polarization, which in turn induce electric waves with analogous oscillating electric polarization – and vice versa. This model of mutual induction should explain the perpetual propagation of light with constant velocity in the aether gas. Thus, the aether waves represent polarization processes of a dielectric (non-conductive) substance. The other idea MAXWELL had in mind was that the aether gas – or the electromagnetic field – can be magnetically polarized just as ordinary gases, i.e., may also form a 'molecular' structure with magnetic dipole properties. So, the emergence of dipolar magnetic properties can be understood as molecular structure formation as soon as we accept that *magnetic* induction causes holistic division and branching processes both in ordinary gases as well as in the aether gas. And if the aether should not exist, it can be only dimagnetic branching processes of single light rays or waves, or more generally, of 'free' electromagnetic fields. Thus, with a holistic interpretation of the division process, it would have been easier to understand the unity of light and matter. But HERTZ and HEAVISIDE had already banished this possibility of thinking from aether physics, which made it – at least theoretically – impossible to recognize physical branching processes behind magnetic polarization. This is the typical effect of 'theory glasses' through which one views the world.

Thus, an possible explanation of the FARADAY and ZEEMAN effects could also be given by holistic division and branching processes, if we combine the dimagnetic polarization of the aether gas, as MAXWELL had in mind, with AVOGADRO's hypothesis of the molecular divisibility of gases, supplemented by the hypothesis that these processes are scale-less and do not depend on energy or matter density. However, MAXWELL's interpretation was itself ambiguous: On the one hand, he conveyed the impression that the aether resembles an incompressible electric fluid; on the other hand, he conveyed the impression that aether waves are magnetic in nature and can assume magnetic dipole properties. HERTZ had moved from HELMHOLTZ's treatment of electrodynamics to MAXWELL's field theory and therefore preferred an electrical view point. For this reason, he probably also ignored the dimagnetic polarization of the aether gas proposed by MAXWELL – or of the "free" electromagnetic field, as MAXWELL used to call the aether from 1864 on. In this way, HERTZ discarded the most important physical idea, which had to be laboriously rediscovered in quantum physics as *magnetic spin*.

So if the excitation of the aether gas, which we call light, is magnetic in nature and divides or branches holistically under the influence of a strong magnetic field, one could also say that here a magnetic field divides holistically. Through this process, an enantiomorphic field is created, that is, a holistically divided field structure which is magnetically oppositely polarized in itself. The effect is that of a magnetic dipole. The magnetic dipole property of gas molecules is well known and could be explained in the same way. However, the idea that polarization effects could be physical branching processes had not yet been born at that time. Therefore, a mechanical separation of light rays was assumed and diffraction was explained accordingly. In order for a light beam to branch and polarize, it must contain several partial waves – at least two. On this premise FRESNEL had built the transversal wave theory of light. The presumed partial rays in a beam of light are deflected and diffracted differently according to their different wavelengths. However, the branching of a light beam, which manifests itself in the doubling of the emission spectral lines, changes the wavelength and frequency of the two partial rays in mirror symmetry. This does not fit with the composite-hypothesis. In addition, each partial ray can be further branched. The same splitting and branching behavior is also exhibited by gaseous molecular matter, which explains the doubling and multiplication of the spectral lines in the absorption spectrum.

The question of the true nature of the division process underlying both the FARADAY rotation and the and ZEEMAN effect is crucial to our ideas about the structure of light and matter. As we have seen, both phenomena could also be explained by physical branching processes. The structural changes on one side would then have to do with the structural changes on the other side, which could explain the interactions of light with matter. However, the ZEEMAN effect has never been considered from this perspective. It was much more obvious to try to combine the particle ideas of Newtonian mechanics with MAXWELL's electromagnetic theory and ignore the contradictory nature of the aether gas for the time being.

The interpretation of the magnetic splitting of spectral lines given shortly thereafter, which postulates smaller particles with negative charge in the atom, has contributed significantly to the fact that the electromagnetic world view could not develop into a conclusive field structure theory of matter. Instead, atomistic particle hypotheses made the race, which seemed to work well in explaining the structure of matter – at least for two decades. Initially, the magnetic splitting of spectral lines could be explained quite well by particle models and motion concepts in the sense of mechanics, at least the occurrence of doublets and triplets. The other cases (with four and more splitted spectral lines), as it turned out twenty-five years later, could be explained only if one regarded the electron particle at least theoretically as divisible and assigned to it a *holistically divisible* wave function which described a *branched electron wave* with exactly opposite magnetic properties. However, the physical meaning of this wave model was not understood. Until today one tries in vain to understand this *dimagnetic polarization* or magnetic dipole formation with rotating electron particles as 'spin'.

Lorentz: The attempt to separate nature into matter and fields

Since gases were assumed to consist of particle-like molecules and these presumably of atoms, one tried to understand the splitting of the spectral lines as a consequence of the interaction between aether waves (light) and electric particles. It is interesting to know how this idea came about in the first place, since all our modern conceptions about the structure of matter have been built on this fundamental assumption. From the introduction of the transverse wave model of light by FRESNEL in 1823 to the theory of MAXWELL and the experiments of HERTZ, a rough, preliminary picture had been imagined: transversal aether waves are made purely mechanically by molecules (mass particles) rotating back and forth in the aether medium. So imagine a stirring rod rotating back and forth, to which a sphere is attached, which sets a jelly- like substance into rapid transverse oscillations (see Fig. 29, p. 204). Since these vibrations are very fast and propagate very fast – after all, these are supposed to be light waves – the jelly would not only have to be highly transparent, but also more elastic and stiffer than steel (otherwise the transverse waves would break off or be damped very quickly as in a viscous dough). At the same time, however, the aether matter would have to be as limp as a sewing thread that refuses to transmit longitudinal pressure forces. These contradictions, which quickly came to light after FRESNEL began to explain the branching and polarization of light beams with transverse waves and the composite- hypothesis, prompted YOUNG as early as 1823 to make the sarcastic remark that FRESNEL's aether, although very ingeniously designed, would have to be *"not only highly transparent and highly elastic, but absolutely solid"*. That was, of course, impossible. How should the planets pass such a medium without any resistance? YOUNG had another idea: There are no solids at all! Ordinary matter would rather resemble coarse nets or structures through which the aether can flow freely, *„perhaps as freely as the wind passes through a grove of trees."*

This picture seems to be worth discussing again, because the modern atom is supposed to be nearly empty – at least if you imagine that electrons circle around an atomic nucleus like planets around the sun. Although we know since hundred years that this picture is wrong, it is still taught in school. Thus, from the very beginning, the transverse wave model encountered insurmountable logical difficulties that made it impossible to understand the elastic nature of the aether gas and the dynamics of 'field structure formation' that Maxwell had identified as electric and magnetic polarization processes. Nevertheless, Fresnel had succeeded in explaining a large number of optical phenomena on the basis of this assumption and modeling them with mathematical precision.

After the experiments of Hertz, the emission of transverse aether waves was attributed to extremely fast oscillating opposite electric polarization states which were interpreted with Maxwell's concept as unidirectional movements of negative charges. In any case, it is correct that the electrical polarization state must oscillate between two opposite states in order to emit light. Hertz had achieved this by a fast oscillating alternating current in a conductor interrupted by a spark gap. So this transmitter acted like a huge oscillating electric dipole, which emitted radiation with certain wavelengths and frequencies depending on the length of the metal rods, the size of the capacitors, and the frequency of the alternating current (which had to be brought into resonance).

In principle, the elementary process of light emission by atoms or molecules would have to work in the same way. Atoms or molecules would therefore have to embody fast oscillating electric dipoles and could possibly even resemble tiny dipole antenna transmitters. The diatomic molecule already comes close to this idea – it would only have to oscillate electrically somehow. However, the fact that Maxwell's unidirectional motion of negative charges was only a provisional model introduced to be able to treat polarization phenomena within the framework of Newtonian mechanics was apparently almost forgotten by the end of the century. Moreover, if light waves were caused by the motion of such charged particles, the frequency of the light would have to correspond directly to the frequency of the mechanical oscillation. But this was not the case, especially when the absorption lines of the molecule split under the influence of extern magnetic fields, which changed the frequencies of the partial rays. So there had to be another, unknown movement, which superimposed the original displacement movement of the negative charges. What kind of movement could it be?

The change in frequency of the split or branched partial rays in the Faraday glass sample and in Zeeman's gases can only be attributed to a change in wavelength, because according to Fresnel's theory the speed of light does not change – if the diffraction of the light beam takes place in the *same* medium. Hence the distance between the wave crests must change, which leads to a change of the *wave period,* i.e., the waves arrive at shorter or longer time intervals at the absorption point. This is reminiscent of the Doppler effect, which, however, requires a relative motion between transmitter and receiver. This is not the case in the experiments of Zeeman. To explain the change in wave period, Zeeman referred to the extended aether theory that his colleague Lorentz had been working on since 1892.

Hendrick Antoon Lorentz had been a professor of theoretical physics at the University of Leyden, Netherlands, since 1877 (when he was just 24 years old). Between 1892 and 1895, he had already developed equations for the motion of positive charged molecules under the influence of magnetic fields, which also took into account accelerations and velocity dependencies that Wilhelm Weber had already considered four decades earlier in his bidirectional electric current model. In this way Lorentz founded the theory of moving charged particles in electric and magnetic fields which became later part of electron theory. With this approach he tried to avoid the debate about the true nature of the aether gas by ignoring, like Maxwell, the contradiction between the elasticity and the strange rigidity demanded by the transverse wave theory.

By the way, this artificially demanded rigidity of the aether immediately dissolves into thin air, if the transversal wave theory can be replaced by something else. Then it becomes clear that this rigidity implied arbitrarily far-reaching, lossless energy transfers of the light (until an absorption event takes place). In this way, the transverse wave concept also obscures that non-local, i.e. global, holistic effects and scale-less natural phenomena could exist. They make sense in a bidirectional interpretation of the speed of light, if one assumes that the structure *formation* of light – the branching – needs time. This is exactly the duration which Maxwell interpreted as the velocity of light, but also considered as a constant ratio expressing the *elastic-*

ity of the aether gas. This elasticity is now hidden behind the material parameters of electric permittivity and magnetic permeability (and susceptibility). LORENTZ followed this path and made the motion of electric charges in the atom responsible for the emission of light. In this way, he attempted to combine FRESNEL's transverse wave theory with the electromagnetic field theory while retaining the aether concept and the atomistic body concept of Newtonian mechanics. As LORENTZ summed his program up in his Nobel Lecture in 1902:

„The theory of which I am going to give an account represents the physical world as consisting of three separate things, composed of three types of building material: first ordinary tangible or ponderable matter, second electrons, and third ether.“

Originally LORENTZ had suspected that positive charged molecule ions, which already embodied the motion of matter in FARADAY's electrolytes, are the origin of the light waves. However, ZEEMAN's experiments had shown that the spacing between the split spectral lines was much larger than expected – and the appearance of quadruplets and quintuplets was an even bigger surprise. So LORENTZ concluded that the mass of the radiation-producing electric particle had to be about a thousand times smaller than originally calculated (or the magnetic force correspondingly stronger, but he did not take this into account). Therefore he recalculated the mass/charge ratio and interpreted the *negative charge* now as a particle with tiny mass. Since the operation of HERTZ' transmitter, the electric dipole oscillator, was explained by unidirectionally moving negative charges, which move back and forth very quickly, he kept this concept of motion: if the atom were a sphere and such a transmitter, one would only have to design a spherical electric dipole oscillator with a periodic motion. So he imagined negative charged particles orbiting in the atom. The speed-dependent frequency of the orbital period then corresponds directly to the frequency of the emitted light waves.

Since moving negative charges induce polar (directional) magnetic fields (in this way MAXWELL had explained the nature of electric current and OERSTED's circular magnetic field around current-carrying conductors), the orbit of the negative particle can be affected by external magnetic fields – and therewith the wave period of the emitted aether waves. Of course, a closed circular orbit was not really the same as HERTZ's interrupted dipole transmitter, but the effect should be the same. ZEEMAN described the development as follows:

„A real explanation of the magnetic change of the period seemed to me to follow from Prof. Lorentz's theory. In this theory it is assumed that in all bodies small electrically charged particles with a definite mass are present, that all electric phenomena are dependent upon the configuration and motion of these 'ions', and that light-vibrations are vibrations of these ions. Then the charge, configuration, and motion of the ions completely determine the state of the ether. The said ion, moving in a magnetic field, experiences mechanical forces of the kind above mentioned, and these must explain the variation of the period." [32]

Recall that the real reason for these considerations was the fact that the splitting of the spectral lines in the emission spectrum indicates a branching of light beams. Correspondingly, the splitting of the spectral lines in the absorption spectrums indicates a branching of the molecules. This applies both to molecular gases, which according to the theory should consist of 'double atoms', and to vaporized substances, which should consist of single atoms and form 'monatomic' gases. The branching of the emitted light leads to changed frequencies, wave periods and polarities in the partial rays, and the branching of the light- absorbing atoms or molecules changes their structure in a mirror-symmetric way. Obviously, the magnetic splitting or branching produced a new physical degree of freedom in the gas. As usual in the kinetic theory of mechanics, one tried to justify, or at least to model, the physical degrees of freedom with possibilities of motion of bodies in space. In this view, degrees of freedom are associated with the dimensions of space (three for translational motions and more for rotary movements).

The splitting of the spectral lines could not be explained by orbiting negative particles, since this motion should already generate the

32 P. Zeeman: "On the Influence of Magnetism on the Nature of the Light Emitted by a Substance" ,1897. He refers to two works by A. Lorentz: "La Theorie electromagnetique de Maxwell", Leyden, 1892; and "Versuch einer Theorie der electrischen und optischen Erscheinungen in bewegten Körpern", Leyden, 1895).

original spectral line. Consequently, a new degree of freedom had to be added: under the influence of an external magnetic field, the circling negative charged particle had to be able to perform yet another movement. This gave LORENTZ the idea that the circling motion is superimposed by a tumbling motion of the negative particle caused by the external magnetic field. This was quite convenient, since one could now apply mechanical analogies and equations of motion as known from coupled pendulums and gyroscopic precession (as early as 1854, KELVIN had proposed a double pendulum model to explain Faraday rotation. Today we know that this is nonlinear dynamics).

ZEEMAN's experiments made it clear that magnetic fields must cause structural changes in gaseous matter. So if the atomic interpretation of the molecule was correct, these structural changes had to occur in every single atom. However, it was not yet clear whether atoms really existed. Even with the best microscopes it was impossible to prove their existence. Consequently, physicists could not yet be sure whether the atom hypothesis, the assumption of indivisibility, was correct at all. And the question of whether the division of molecules in chemical reactions is compatible with the concept of homogeneous matter that can structure itself also remained unresolved. For this, one would only have to develop suitable physical- topological ideas of how gases can structure themselves through holistic division processes. That AVOGADRO's molecular divisibility can actually be interpreted in this sense could not yet be recognized in the 19th century, since the concept of cell division was limited exclusively to biology and developed only slowly (see p. 77). By the way, one physicist who was quite aware of this possibility and had already tried such approaches was KELVIN. However, he did not start from AVOGADRO's hypothesis of molecular divisibility, but worked abstractly mathematically on topological knot theories. By the way, holistic divisibility could also explain the influence of temperature on the structure of matter, leading to a new kinetic theory of heat.

So the splitting of spectral lines had raised the question of how the structural changes of matter are related to the properties of light and how this is to be understood in the context of electromagnetic aether and field theory: As holistic, bidirectional charge generation and separation (as polarization in the original sense, i.e., as dielectric polarization by branching), as bidirectional motion of opposite half-charges in the sense of WEBER, as unidirectional motion of exclusively negative charges in the sense of MAXWELL, or as dimagnetic polarization (aka magnetic dipole formation) of the aether gas. This is the same fork in the path of cognition at which YOUNG, FRESNEL, FARADAY and MAXWELL have already stood. Obviously, the question is which philosophy of nature is the more appropriate: an atomistic-mechanistic world view or a holistic, structural understanding of nature.

LORENTZ' hypothesis and design concept – that the atom contains circular moving negative charged mass particles – was shortly thereafter seemingly unambiguously confirmed by the theory of cathode rays proposed by J. J. THOMSON. THOMSON's theory stated, experimentally well founded, that cathode rays consist of tiny mass particles carrying a negative elementary charge. This was considered as the discovery of the electron – the first subatomic particle – and was honored with the Nobel Prize in 1906. Thus the existence of electron particles became a seemingly unambiguously proven fact.

Experimentally unambiguously proved, however, is the fact that the double-slit experiment disproves both the particle interpretation of the electron (1) and the assumption of its indivisibility (2). In 1906, nobody could know yet that these two assumptions would fail in this experiment, but one could have suspected it already – because the same problem occurred already in 1905 with EINSTEIN's light quantum hypothesis. Since 1925 we know from diffraction experiments on crystals that electron beams behave like waves, which means that electron beams must be divisible like light beams and cannot consist of particles. This led to the theoretical acceptance of wave equations for electrons, but was not really understood physically. In 1926 it became clear that this must also be true for atoms. What this means for the constitution of Nature and reality could not be clarified to this day (in the meantime we write the year 2023). So there must be another explanation for the nature of electron rays and the structure of molecules (and light). That transversal wave theories are not sufficient for this, we know already from the quantum theory of light and matter. This is exactly the task that a new generation of physicists will have to solve as soon as Sleeping Beauty wakes up from her 100-year deep sleep... So let's have a look how THOMSON came to his theory and how it can be possible that the particle interpretation is wrong although the theory describes the phenomena quite correctly.

Thomson: The invention of the electron

In 1897 J. J. THOMSON (1856-1940) attempted to explain the quantum properties of cathode rays, which LENARD had discovered ten years earlier, by the existence of electrically negatively charged particles. It is often said that he *discovered* the electron, but the truth is that he assumed from the outset that cathode rays consist of negative charged corpuscles. What THOMSON actually did was to design a constructive theory based on a particular hypothesis. If the theory is compatible with experiment, the underlying hypothesis can be considered provisionally confirmed. However, if new experimental facts come to light that disprove the original premise, even a constructive theory that is internally consistent may prove untenable. In this case it is the double-slit experiment with single electrons, which only in 1959 could finally prove that the interpretation that cathode rays consist of particles must have been wrong.

It is quite understandable that this was (and is) not easy to accept, because we also know very well that the electron theory is an almost perfect model of the behavior of cathode rays. Electron beams are used for many purposes. For many decades, television sets worked perfectly on the basis of cathode ray tubes. Electron microscopes also work on the basis of this theory – consequently, it cannot have been completely wrong. Moreover, the entire semiconductor technology is based on the electron-particle concept (supplemented by quantum mechanical subtleties), which led to the rapid rise of the computer and software industry in the second half of the 20th century. But we also know that the particle idea must be wrong, because strictly speaking they are only elementary quanta *of action* of negative electricity, which basically behaves like 'electrically non-neutral' light of short wavelengths and can be divided in diffraction processes. The electron microscope operates on the principle of diffraction, or, to put it bluntly, it relies directly on branching processes just like any ordinary optical device. So far, such processes can only be modeled with wave theory, although the wave model is not able to reproduce the quantum-like behavior that occurs in the local energy transfer of light and electron radiation. This is the same problem as with the nature of light. This is again the tricky quantum puzzle, and again we note that there must be another way to understand these natural phenomena. So let us look at THOMSON's electron model:

THOMSON explained the deflection of cathode rays by electric and magnetic fields as a deformation of the trajectory of moving electron particles, which must therefore also have magnetic properties. Thus the electric ray must be surrounded by a polarized magnetic field, like a current-carrying wire. Based on certain considerations and analogies (speed of the particle, ballistics, inertia of moving mechanical bodies), he calculated the mass of the hypothetical particle from the curvature of the electron beam in electric and magnetic fields, so that the *particle* concept (1), the *indivisibility* of the particle (2), and the idea of an electron *trajectory* (3) suddenly seemed very convincing. But exactly these three concepts fail at the double slit.

The mass of the imagined electron corpuscle was 1800 times smaller than the mass of the hydrogen atom, set equal to one as the unit of comparison. This atom is equivalent to AVOGADRO's elementary molecule, but the integral molecule, the *potentially* divisible molecule, would have the mass unit two. Relative to the molecular mass, the electron mass would therefore be only half. THOMSON was free to choose the sign of the charge by convention; he chose a negative one. And since the origin of the released electric beams should be in the atoms or integral molecules of the cathode material, the components remaining in the material must be positively charged atoms (positive molecule ions). This in turn immediately led to the insight that the so-called atom, *the indivisible*, could not really be indivisible. Thus, the term atom was actually wrong. Nevertheless, the term atom was continued to be used in the sense of an *elementary chemical unit of matter*, because in this case, the atom or integral molecule was not really split in two halves: figuratively speaking, the apple was not divided, but only peeled... After all, the atom remains as a positively charged ion, it only loses its negative charge (its shell, so to speak) and thus only an insignificant part of its mass.

With this work it became clear that LORENTZ' negative electric particles were the same as THOMSON's electrons. This gave a boost to the supporters of the atomic world view, although the atom could now no longer be the smallest building block of matter, as it seemed to be made up of even smaller 'atoms'. In this way the atom hypothesis turned into an atomistic hypothesis. The idea of an electron particle, in turn, led automatically to the conclusion that atoms must be composed of mass particles with opposite electric charges, since atoms are normally electrically neutral. The idea that these charges are not present from the outset, but could be created by charge separation, i.e., by dielectric polarization processes, understood as field structure changes by branching, was not considered. Presumably, the mechanical and reductionist idea of an additive composition of matter was so attractive that it steered scientific research activities in a new direction: suddenly it seemed possible, if not mandatory, to combine Newtonian mechanics and the atomistic world view with electromagnetic field theory.

Through LORENTZ and J. J. THOMSON works, the conception developed that accelerations of electrons, abstracted as point masses with negative charge, cause the emission of light. This was a rather foggy idea, since it turned out that the electron could be conceived neither as a volume element nor as a dimensionless point. How should a point-like electron without any geometrical dimensions be able to produce light? Where comes the light and its energy from? As one could see from the experiments of HERTZ, TESLA and many others, the emission of light had to do with the periodic change between two opposite electrical polarization states, but also with electrical discharges. Where were they in the atom? How should LORENTZ' closed electron orbit accomplish electric discharges? All these questions make clear that the theory of light emission must be directly connected with the theory of induction and electric current.

The fact that there was no reasonable answer to the question how individual electrons without dimensions should be able to produce light was not hung on the big bell, because the goal of physics had changed in the meantime: physicists now tried to explain the structure of matter with smaller indivisible particles, ascribing properties such as mass, spatial extension, speed and charge to them, and combining atomistic ideas of the constitution of matter with the wave theory of light. At first it almost seemed as if LORENTZ had really succeeded in establishing a kind of mirror symmetry between matter and fields (particles and aether waves) with his mathematical model, even if they first had to be separated in the mind for this purpose. But as it turned out, the equations could not be correct, as beautiful as they appeared, because the revised MAXWELL-HERTZ-LORENTZ equations failed miserably at the point where the electric fields were supposed to have their origin: at the electron. Because at a mathematical singularity, a dimensionless point, the electric field strength becomes infinitely large. This makes no physical sense and cannot be observed. Obviously, there was something fundamentally wrong with the idea that electron corpuscles are the source of MAXWELL's electric fields. However, nobody seriously considered that the particle idea could be wrong on principle. But if electric fields are the product of dipolarization processes – of a separation into negative and positive half-charges – we are dealing with holistic division and branching processes and bidirectional motions.

Similar problems occur again and again with dimensionless points, just think of the *big bang* or the *black hole singularity* in the theory of gravitation. In the mid 1980s some mathematical physicists tried to avoid singularities by introducing two points connected by 'strings'. However, to date they have not been able to physically substantiate this concept of a *mathematical tool* or provide experimental proof of the existence of strings, which were originally regarded as new elementary units to replace particle concepts. This is probably because mathematical speculations can falter very quickly if they cannot be based on an experimentally founded physical principle. Perhaps our field branching principle can provide new insights here as well. In 1906, however, most physicists did not regard the problem of the electron as an artifact, which only appeared with the introduction of the body concept of mechanics and unidirectional motion into field theory. They saw the problem rather as a technical question of model building: couldn't one simply consider the point- like electron as a three-dimensional volume or matter unit in order to avoid this annoying problem? Then the electric field just starts at the surface – which would solve this problem. However, this immediately led to another problem: How then to distinguish the particle's own electric field from external electric fields? This is the secret of the electron, which still occupies us today.

In 1924 PAUL A. DIRAC designed a mirror-symmetrical equation of opposite electric charges and predicted the existence of an antiparticle of the electron with equal mass but opposite electric charge, later called *positron*. In this way, DIRAC almost rediscovered the concept of a bidirectional motion of opposite charges, which already implied a time symmetry. However, he was not aware of this relationship, because he tried to explain the electron as an empty place in a *sea of positive charges* (in contrast, in 1864 MAXWELL had explained the positive charges as absence of negative charges). And in the 1940s, a young physicist named FEYNMAN began again to seriously study the inherent *time symmetry* of the MAXWELL- LORENTZ equations, which was well known mathematically but had received little physical attention until then. This time symmetry also expresses a mirror symmetry which states that, literally, the physical processes involved (electric & magnetic polarizations) are reversible and the motion of light is *bidirectional.* Many physicists believe that the time symmetry of the MAXWELL equations is a mathematical artifact resulting from the use of imaginary numbers that have no physical meaning at all. Although FEYNMAN could not really solve the time symmetry problem, he explored the consequences for a new understanding of the interaction of light with electrons. As it turned out, one would either have to abandon the concept of time (that is, declare it physically senseless), or allow backward running physical effects in time. In other words, one would be forced to accept a mirror symmetry in time (a *reversibility* of all physical processes), in opposite electric charges, electric currents, electric current pulses, and in the direction of propagation of light. Doesn't this point directly to bidirectional motion and enantiomorphic fields?

The idea of LORENTZ and THOMSON to combine the atomistic particle model of the electron with the electromagnetic field or aether theory led to irresolvable contradictions in the understanding of the true constitution of matter and light, the full extent of which, however, was to become clear only years later. Initially, their work led to the idea, still widely held today, that cathode rays, light-induced electric radiation, and electric currents consist of a stream of particle-like electrons flying through space like tiny billiard balls. This conception was later also applied to light quanta and their motion. But the philosophical assumption that there are indivisible objects in Nature was in direct contradiction to the interference effects and wave theory, which had seemed so promising until then. One reason for this fateful development was that the constitution of cathode rays and light-induced electric radiation could not be explained by the transverse wave model, whereas the particle model could at least reflect the quantum properties of the energetic action. Wave theory was thus either incomplete or simply not applicable to matter – or so it was thought at least at that time. Responsible for this was the assumption that light has no mass, is always electrically neutral and is continuously transferred to matter. But how light really transfers its energy to matter was by no means clear. There was practically only guesswork, since neither the true physical constitution of matter nor of the light was known.

The photoelectric effect and the quantum properties of cathode rays had shown that the assumption that rays continuously transfer their energy to matter could not be correct. This was true not only for electron beams, but also for light, as was soon to become apparent. One idea (pursued by TESLA, among others) was that electric rays could be longitudinal waves that act only in a way that can be confused with particles, since they act locally and transmit an impulse to matter (kinetic energy, called 'momentum'), which in mechanics describes the impact of a body. However, physicists who wanted to explain the world on the basis of electromagnetic principles insisted that the radiation character of electric rays cannot be ignored and that the mechanistic-atomistic interpretation was wrong. That they were right did not became clear until 1923-25, when it was discovered that the properties of the electron in the hydrogen atom may also be modeled by a wave theory, that electron beams can indeed be diffracted (branched), and that electrons must be magnetically polarizable. However, the whole drama became very acute only in 1927, when it finally became clear that even single atoms, electrons and all other 'elementary' particles, which had been invented in the meantime (protons, neutrons), should be at least theoretically divisible and therefore could be modeled only by wave theories. This is the so-called *wave character of matter*, which is valid for molecules, atoms and electrons and all elementary particles. This term expresses practically the theoretically not yet understood holistic divisibility of supposedly indivisible particles.

The initial successes of ZEEMAN, LORENTZ and THOMSON led first to the integration of the electron particle concept into electromagnetic field theory, then to the theory of special relativity, then to the planetary model of the hydrogen atom, and finally to the model of a standing wave of negative electricity, which describes a spherical shell of the atom, which can structure itself by the absorption of light, creating the opposite magnetic spin states of the electrons. As we will see in the second volume, this process practically describes the formation of a hierarchically branched magnetic field structure.

This makes the ZEEMAN effect – the magnetic branching of molecules and electromagnetic fields – a key experiment in physics and LORENTZ's theory a linchpin for understanding the roots of special relativity and subsequent developments of atomic physics. However, the fact that the particle idea and the motion concept of mechanics cannot reasonably explain the structure of atoms and molecules should become clear only in 1926, leading directly to the interpretation problems of quantum physics: if even electrons and atoms are divisible at the double slit, the particle notion must be plain wrong. The fact that one could also be dealing with branching processes of magnetic fields was not considered at all – the idea was still not in the air.

According to the rather arbitrary definition of HERTZ, the plane of the "electric wave" is appointed as the reference plane of polarization. But actually, as we have seen, this wave should be of magnetic nature, because its wave curve describes only the electric effects (the intensity of spark discharges), but not their cause: according to FARADAY's induction principle, the cause of the sparks can only be a changing magnetic field at the location of the hand-held ring detector, which induces a voltage in it. Consequently, we can just as well imagine these waves – and light in general – as magnetic fields with a dipolar structure. In fact, polarization is always a *magnetic property*, even in electrons, atoms and molecules. So we have found a commonality in the constitution of light and matter, which now allows us to attribute the structure formation of nature to magnetic fields that can divide holistically.

Thus, the goal is to understand the magnetic branched structure of light and matter behind the transverse wave model, diffraction phenomena, and stationary wave models, for which, strangely enough, no time passes. The magnetic induced twist of the chirality axis of the branched ray system in FARADAY's rotation experiment actually takes place in reality, but this cannot be considered as a proof of the transversal wave model. What can be said with certainty, however, is that the acting longitudinal magnetic field in the experiments of FARADAY and ZEEMAN must cause an anisotropic structural change in gaseous matter, a magnetic dipolarisation of the entire molecular structure, leading to the division of the incident light ray into two left- and right-handed magnetically polarized partial rays. This characterizes primarily an ontological mirror symmetry, a chirality.

And exactly this indicates that we have to do here with a branching process, because with a mechanical separation of two light rays this antisymmetry cannot be explained. In addition, the frequencies of the partial rays change mirror symmetrically by almost exactly the same amount. In these facts we can recognize an enantiomorphic, mirror-symmetrical magnetic field. This shows that our branching hypothesis is quite compatible with the experiments. However, this is only circumstantial evidence for a holistic division process. For a real proof, we still have to show that this also applies to single light rays and molecules, that the two partial rays or wave parts are coherent, i.e. form a physical unit and act energetically always as a whole. This is exactly what quantum physical experiments will prove.

All this also fits well with our holistic interpretation of the molecule division process: a polarized light ray – representing a polar magnetic field branch – induces a molecular cell division of integral molecules (our atoms), which causes the opposite magnetic polarization of the two daughter cells. Integral molecules thus become 'double atoms' and magnetic dipoles only by holistic division or branching processes, which explains the magnetic anisotropy of the gas. That this interpretation represents a new point of view (as far as I know), I must emphasize here once more. I ask the protesting skeptic for some patience, but refer already here to the STERN-GERLACH experiment, where beams of originally *non-magnetic* silver atoms branch at inhomogeneous magnetic fields, to double-slit experiments with single electrons, atoms and molecules, and once again to the message of the molecule orbital theory. We will discuss all these experiments in detail in the second volume.

Lorentz, Fresnel and the mystery of relative motion

LORENTZ's approach to the unsolved problems of aether physics was a historical turning point: it was the first time that anyone separated fields from matter. Such a separation did not exist in FARADAY's and MAXWELL's thinking, since they assumed – despite certain differences – that gaseous matter and fields are basically the same thing. Paradoxically, it was the structural properties of the cathode, beta and alpha rays that seemed to indicate discrete particle-like actions and energy transfers. Were they possibly polarized partial rays of a branched field, just as the incident light ray in the FARADAY rotation experiment? This seems to be an idea that no one has pursued yet: cathode rays are then electrically negative polarized rays, their counterpart would be electrically positive polarized rays. Alpha rays are then electrically positive polarized helium molecules. In our new interpretation of molecular divisibility, they would be the cellular components of an expanding gas volume with a certain kinetic energy and density. The counterpart would then be electrically negative polarized electron rays. Both types of radiation behaved in principle like longitudinal (gas) waves. They could not yet be interpreted as polarized partial waves, at least not at that time, because the idea that individual electrons, atoms and molecules behave like gas jets (with the lowest possible density) that can be holistically divided or branched simply did not yet exist. But if the branching hypothesis is correct, i.e. physical branching processes really exist (independent of the density of the gas or the energy density of the light), then we could also speak of a longitudinal wave which branches mirror-symmetrically. Because the quantum and pulse-like energy transfer by the branched partial rays resembles the effect of longitudinal waves, although the two branches can transfer their energy (according to the quantum theory) only alternately. From this follows, if the whole emitted energy is transferred at only one point, that the branched structure must recede during the energy transfer. Could this explain the problems in the interpretation of the MICHELSON-MORLEY experiment and the FRESNEL coefficient?

The insufficient understanding of the basic assumptions of molecular theory, optics and electromagnetic theory apparently led to the fact that physical branching processes were not considered at all and thus their discovery was missed. Without this structural principle, it proved impossible to design a unified electromagnetic world view for light and matter. Instead, the atomistic-mechanistic world view prevailed. Within twenty-five years, it became the dominant paradigm, until the incomprehensible failure of the indivisibility hypothesis was encountered. Although in 1927 it became clear that the superposition postulate was indispensable, had to be valid also for electrons and atoms, implied holistic divisibility and thus practically described the failure of the atomic paradigm, the failure of the atom hypothesis was neither understood nor accepted as a physical reality. The unrecognized failure of the atomic hypothesis, the not understood nature of polarization and the attempt to separate matter and fields were and are the reasons that prevent a genuine understanding of the true constitution of matter and light.

At that time, however, it seemed that MAXWELL's field theory could model the interaction of magnetic fields and electric charges at least mathematically sufficiently well. That MAXWELL's treatment of the induction principle could not truly reflect the symmetry (relativity) of motion between the magnet and the conductor only later proved to be part of the problem that EINSTEIN was the first to recognize. However, since the explanation of the magnetic branching of light waves *"through a modified motion of 'ions'"* was considered *"correct or at least highly probable"* (ZEEMAN), almost all physicists remained trapped in particle conceptions and mechanical models of motion. LORENTZ idea that circling electrons produce the aether waves was an adaptation of MAXWELL's idea that unidirectional moving negative charges induce magnetic fields *and* light waves. Once again we note that, according to FARADAY's induction principle, light should actually be *magnetic* in nature.

LORENTZ had to take note that the nature of light poses complicated riddles, which had to do with the constitution of the aether, FRESNEL's transverse wave concept (the nature of polarization) and the definition of the speed of light. On the one hand, there was the experiment of 1851, in which FIZEAU had investigated the influence of flowing water on the speed of light. He had found that the propagation of light with the flow of water decreases the speed of light, while the propagation against the flow of water increases the speed of light. It is noteworthy that also in this case we are dealing with branched partial rays which develop different wave periods and thus different speeds relative to the liquid medium, just as in the DOPPLER, FARADAY and ZEEMAN effects. The extent of the changes in wave period due to the relative motion of light against a medium has been accurately specified by the FRESNEL coefficient, also called *aether entrainment* coefficient. This coefficient now acquired a new meaning as a correction factor for the speed of light, both in relation to the MICHELSON-MORLEY experiment and to the ZEEMAN effect.

The idea that *moving* translucent matter binds and entrains a tiny fraction of aether gas was originally introduced by FRESNEL in 1816 to make *stellar aberration* consistent with the theory of refraction, which at that time was still based on the longitudinal wave model (we will come to this in the next section). Basically, FRESNEL was concerned with preserving the validity of the theory of refraction for both stationary and moving optical systems. This was the first attempt to maintain the principle of relative motion in optics, which seems violated by the aberration problem. If the laws of optics apply whether the optical system is moving or at rest, the aether can be thought of as an ordinary gas filling a container. Then the telescope moves with the Earth around the sun through an aether gas filling a volume whose size is unknown (this is NEWTON's absolute space, the aquarium, i.e. the cosmos). On this path, it moves sideways to the incident starlight twice a year, sometimes to the left and sometimes to the right. And twice a year it moves in the direction of the line of sight, moving once toward the star and the other time away from it. Since the light needs time for its propagation, this would have to affect the wave period accordingly. While this was easy to explain for the motion of the telescope in the line of sight to the star (with the DOPPLER effect, which, however, had not yet been formulated), the lateral, transversal, or sideways motion of the telescope with the Earth apparently had no effect, since the refraction of the incident light never changed. Exactly this fact FRESNEL tried to explain with his aether entrainment hypothesis. However, having failed to explain the phenomenon of polarization with the longitudinal wave model, he introduced the transverse wave model in 1823. Accordingly, he had to adapt the aether entrainment hypothesis. It was precisely this hypothesis, now based on the transverse wave theory, that FIZEAU's water experiment twenty-four years after FRESNEL's death was able to confirm: the aether entrainment coefficient indicated a mirror- symmetric deviation from the wave period of the incident light beam resulting from the motion of the partial rays relative to the water flow. And since, according to the theory of refraction, the change of the wave period implies a change of the velocity of light, the speed of the two partial rays must differ by the same amount depending on the direction of flow of the water. FRESNEL's aether entrainment coefficient thus described a mirror-symmetric change in the local speed of light, which followed from the relative motion of the partial rays to the translucent medium.

Although this appears to be a purely kinematic image, FRESNEL suspected a dynamic feedback effect behind it. He assumed that the water interacts with a tiny part of the entrained aether gas, which causes a change in the time cycle of the wave. If I am not mistaken, behind this is a movement of light that relates the distance traveled to a rate of change of the unit of time per unit of time. This is exactly the way acceleration is defined in mechanics. In modern words, FRESNEL assumed that a still unknown interaction of light with matter slows down the propagation of light, which basically describes a kind of friction. If this is true, we are dealing with hidden dynamical processes in which the aether gas – the light – interacts with the water molecules, i.e., changes the molecular structure of the water, and of translucent dielectric matter in general.

In a purely kinematic picture, the relative motion of material bodies of any kind through the aether gas should lead to the same effects. This means that the actually constant velocity of light in the aether appears smaller or larger for a moving body. In FIZEAU's experiment this material body was the liquid water, in the MICHELSON-MORLEY experiment a solid body: the interferometer device mounted on the moving Earth. So if a body moves against the aether waves or a ray of light, the local speed of light appears larger, if it moves with the light wave or light ray, it appears smaller. This is expressed by FRESNEL's entrainment or aether drag coefficient, which now represents a speed correction factor for the relative speed of light: the amount of the deviation is the same in both directions and depends only on the ratio of the body's velocity to the speed of light. In the MICHELSON interference experiment, however, the expected change of the wave period – and therewith a speed deviation between the branched partial rays – was not detectable. One wonders at this point, what the DOPPLER effect has to do with FRESNEL's speed correction factor, since both suddenly appear to be involved. Anyway, the motion of a body relative to the aether or light rays was not detectable with an MICHELSON interferometer. This was puzzling, because FIZEAU had been able to prove this motion relative to the light rays clearly with an interference experiment in water, which he had designed expressly to prove the aether entrainment effect.

So what was going on there? Are there possibly other physical processes that complicate – or prevent – the detection of relative motion of matter through the aether gas or against aether waves? Or is there no relative motion at all, as in the following proposal: Is it possible that an oppositely polarized *standing wave* is generated in the interferometer by branching, which has a fixed branching point in the glass body, which, connected to the interferometer, moves through space together with the branched wave? Is this a 'wave' at all? Is it a bifurcated orbital, i.e. a stationary, spatially branched, but non-oscillating entity? Couldn't this be a branched magnetic field? If we imagine spherical wavefronts instead of light rays, we recognize in this picture a holistically divided 'light molecule' moving through space like an ordinary gas molecule, but expanding further...

Why the DOPPLER theory failed in the MICHELSON experiment was to puzzle physicists for years to come. The actual problem was not easy to understand because the transverse wave theory seemed indispensable to explain the phenomenon of polarization. FRESNEL had linked the material properties of the aether to the refractive index, which MAXWELL successfully related to the electrical and magnetic properties of the respective media. In this way, the elastic properties of the aether – and all other translucent media – were traced back to the electromagnetic properties of matter, which in turn found their expression in the speed of light in the respective medium. According to MAXWELL's understanding, the *constant speed* of light in the aether was a direct consequence of the elasticity of the aether gas, which, as in the acoustic longitudinal wave theory, should originally lead to a compression of the waves in front of the moving body, and to an increase of the wave distances behind it (the waves remaining behind are called retarded waves). The MICHELSON experiment now seemed to show that such a compression effect does not exist in the aether gas.

From this was concluded that the speed of light in the aether must be the same in all directions even for moving bodies, in contrast to ordinary gas theory, where the speed of sound waves depends on the relative motion between transmitter and receiver. This would also mean that the velocity of light in the aether gas is independent of the wavelength, unlike in ordinary translucent media. Remember: In the glass prism or in water, the extent of refraction of light rays depends on their wavelength, which FRESNEL had justified with different (wavelength-dependent) propagation speeds of light rays. In this way the aether got more and more properties which seemed to exist nowhere else in nature. But the MICHELSON-MORLEY experiment failed for *two* reasons: On the one hand, there was no relative motion between light source and receiver (the screen with interference band), on the other hand, one had to deal with an yet unrecognized branching process. This would mean in the aether theory, however, that the aether gas already at the smallest excitation, which we call 'light', divides holistically and polarizes magnetically opposite. Thus the aether gas is not 'immobile' at all, at least not locally. Thus, in addition to the undulation motion describing polarization cycles, there is a structural branching movement which doubles the waves or rays in a specific way. This is exactly the missing movement which WILLIAM THOMSON (KELVIN) had postulated to explain the FARADAY rotation, and the same branching movement explains the splitting of the spectral lines. Does this mean, then, that we can dispense with the notion of subatomic particles if we want to explain the magnetic branching of spectral lines? Can we explain the electrical and magnetic structure formation processes of matter and light much better with a field branching theory?

And of course this branching process has to do with the theory of refraction as well as with the theory of diffraction: In the MICHELSON-MORLEY experiment, the beam splits in the glass body, resulting in a phase shift between the two partial rays of exactly half a wavelength. That is, the glass seems to interact with *only one polarization cycle*, half a wavelength, of the transmitted partial ray. The other is reflected at the surface, at least according to current theory. That is, matter seems to interact with only one partial ray, which is magnetic 'polar'. So the branching point could well be anchored in the glass body by this interaction, from which the two partial rays propagate again with speed of light. Consequently, the two partial rays show no differences in the speed, although they are shifted by a half polarization cycle and can interfere. Has this also something to do with the splitting in the ZEEMAN experiment, where the light beam branches into two partial rays which differ from each other by the same frequency? Is it possibly a matter of uniting the theory of refraction with the theory of diffraction in a new dispersion theory of light?

However, this way of thinking was not pursued. LORENTZ decided to take FRESNEL's speed correction factor seriously, but to ignore the aether gas entrainment interpretation for now. To adapt MAXWELL's theory of light to these facts and to reconcile it with an apparently constant velocity of light in all space directions, he finally integrated ideas which came originally from FITZGERALD and LARMOR, who had proposed *variable* length and time scales. The idea was that a material body is held together by electric and magnetic forces that change as the body moves relative to the aether. This would lead to a shortening of the length scales – and therewith of the body – in the direction of motion (only), which would provide an explanation for the null result of the MICHELSON-MORLEY experiment. Thus, it was a matter of explaining the failure of the Doppler effect in experiment with new auxiliary assumptions and reconciling it with MAXWELL's electromagnetic aether field theory.

In 1889, FITZGERALD had already postulated a contraction of moving bodies in the direction of motion, caused by speed-dependent magnetic attraction forces between atoms or molecules, to explain the null result of the MICHELSON-MORLEY experiment (HEAVISIDE, however, claimed that FITZGERALD had originally assumed a contraction transverse to the direction of motion, which, incidentally, would also fit better to WEBER's bidirectional electric current theory). And in 1897, LARMOR had proposed that time cycles slow down for electrons with increasing velocity, which should be caused by a tumbling motion superimposed on the circular motion of electrons in the atom. This *gyroscopic precession* should lead to a change in the time intervals between recurring positions of the electron, which in turn should determine the wave period of the emitted light waves. So the basic idea was that the change of wavelengths and time intervals of the effective interactions of light, which should be caused by the motion of matter, is exactly compensated by mirror-symmetric changes of length and time scales in the material body. In other words: If light and matter move relatively to each other, the structural changes of the light are compensated mirror-symmetrically by the structural changes of the matter. Details of the atom model construction are not decisive, but must satisfy this principle.

Lorentz first introduced the change of length scales to explain the period changes of the split light waves in the Zeeman experiment, and combined it later with a change of the time scales, attributed to the periods of orbiting and precessing electrons in atoms. With this adjustment he also pursued the intention to explain the stellar aberration and the null result of the Doppler effect without sharing Fresnel's idea of an aether entrainment. However, this meant that the Fresnel speed correction coefficient had to be derived in a different way, which A. Lorentz succeeded in a remarkable way: To ensure the invariance of the velocity of light – or the constant c, whatever its physical meaning may be – he devised a mathematical concept that basically describes the relative motion between matter and light (aether waves) as a mirror-symmetric inversion of the three space coordinates and the time coordinate.

Stellar aberration, aether entrainment effect, and enantiomorphic spaces
Let us conclude the first volume, and with it the 19th century, by reflecting once again on the conceptual puzzles that troubled Fresnel as early as 1818, but were still not solved in 1900. The longitudinal wave theory of Huygens, Young and Fresnel was abandoned in 1823 because the polarization of light could not be integrated into this theory. The idea of the holistic divisibility of the gas molecules and the physical branching of light would have offered the possibility to unify the theory of the aether gas with the ordinary gas theory and thus with the theory of light. The transverse wave theory made this unification impossible, because it required contradictory properties of the aether gas. It also complicated the explanation of stellar aberration, which was already at the beginning of the development of wave theory.

Although stellar aberration is so difficult to understand in the context of the transverse waves that most physicists wisely avoid mentioning it at all (let alone discussing it in a popular science book), we will nevertheless familiarize ourselves with it. This is of interest insofar as it allows us to better understand the cognitive problems that piled up towards the end of the 19th century. Then we can also better comprehend why Einstein's special theory of relativity was perceived by many physicists as a liberating blow, even though it did not yet allow clear ideas about the true nature of reality it seemed to describe. By Einstein's own admission, the problems in explaining stellar aberration and Faraday induction inspired him much more to the principle of relativity than the Michelson-morley experiment. Of particular interest to us, of course, is the question of whether special relativity can be understood as a disguised description of branching processes of light.

Stellar aberration means an apparent change in the positions of stars caused by the motion of the Earth around the sun. However, this effect has nothing to do with the other apparent deviation of the star positions in the firmament which we call *parallax*. Parallax is easy to understand: close your left and right eye alternately and you will notice that the observed object seems to change its position. So parallax is the angle at which a fixed object is seen from two different locations. If the distance between the locations and the angle to the object is known, the distance of the object can be calculated with angular functions. The smaller the measured angle the further away the object is. The same happens when we observe a particular star from opposite points of the Earth's orbit: the star seems to have changed its position, even if the change is very small.

After Nicolaus Copernicus (1473-1543) had introduced his heliocentric world view in 1543, it was recognized quite fast that the circular motion of the Earth around the sun would have to be provable in the form of a star parallax. The largest possible distance between two observation points corresponds in this case to the diameter of the Earth's orbit or twice the radius, i.e., the distance between sun and Earth. If one measures the angle at which the star appears at the opposite points of the Earth's orbit (e.g., March 21 and September 21), the position of the star in the night sky would have to appear shifted a tiny bit. In the course of the year, the star would then have to perform a very small circular or slightly elliptical motion, corresponding to the motion of the Earth around the sun. But for more than one hundred eighty years no astronomer succeeded in proving this parallax effect. Consequently, either the Copernican theory had to be wrong, or the stars were so far away that their parallax appears so small that it is no longer measurable. As it turned out later, the resolution of the telescopes was not yet sufficient.

In 1725, JAMES BRADLEY (1692-1762) began his astronomic observations in London to unambiguously prove stellar parallax. He used a large, newly developed zenith telescope pointing vertically upward (built into an unused chimney). Every month for years, he measured the angles at which the stars appear as they pass over London at zenith. As a measure of great distances he introduced the *Astronomical Unit* (AU), the distance between the sun and the Earth, which amounts to approximately 150 million km. The astronomical unit thus provides a scale of length with which we can express distances in space as multiples of it. What he found out during his observations, however, was surprising: the stars actually describe detectable, almost circular orbits in the course of the year. The effect was even much larger than expected from parallax measurements. And still this effect could have nothing to do with angle measurements indicating different distances of stars, because this angle was the same for all stars. So what had BRADLEY measured with it? Let's consider which possibilities of thinking arise from this fact:

Was PTOLEMY's world view perhaps correct after all? Are the fixed stars all equally distant and attached to a rotating celestial sphere? Or are the stars actually at different distances? Can their light, for some reason, perhaps still reach us at the same time? That the stars are all equally distant seems quite unlikely if they are suns, which in principle are no different from ours. Obviously, these always constant angles are not at all about distances as in the ordinary parallax measurement. But the apparent deviation of the star positions must also depend on the motion of the Earth around the sun, as in the case of the parallax, because the apparent position of the stars circles around a certain point in the course of the year, and this in such a way that the measured angle, which determines the diameter of this circle, is the same for all stars (therefore it is called *aberration constant*). Hence, this apparent change in stellar positions must have other causes. Was the distance between the stars possibly irrelevant for the light? Did "distances" and "time" exist at all in the universe, for light, for the Creator, for God? Or was light just the excitation of a diluted gas which fills the whole space, includes all stars and worlds and therefore always seems to arrive with the same speed? Was the "speed" of the light possibly a constant which had nothing at all to do with movements in the sense of "change of location"? But if it would be so, why does the motion of the Earth around the sun then produce a circular aberration curve – and the same for every star?

After BRADLEY had considered different hypotheses, he came to the conclusion that the apparent stellar motion had to indicate a *velocity difference* between the Earth and the velocity of light. Stellar aberration is therefore not a distance parallax, but a 'velocity parallax'. Since the particle hypothesis of light was still predominant in 1728 (NEWTON had just died), BRADLEY sought an appropriate explanation: if all the particles of light have a constant velocity, the motion of the Earth relative to the incident light particles causes a velocity difference, which is virtually like a *locally varying speed of light*. The Earth moves on its orbit around the sun sometimes in, sometimes against and sometimes across the direction of the incoming starlight, which leads to a regularly varying velocity difference. So the light arrives sometimes earlier and sometimes later, which leads to the shifted star positions. Based on these considerations, BRADLEY calculated the velocity of light to be about 295,000 km/sec from the aberration constant and the mean velocity of the Earth. From this also the famous conclusion followed that the light of the sun needs about eight minutes to reach the Earth. The phenomenon of constant stellar aberration thus also seemed to support the assumption of a constant speed of light in space, which had to be independent of the color or frequency of light. This had a lasting influence on all subsequent theories of light, including FRESNEL's wave theory.

In contrast, the aberration constant cannot be explained by different velocities of the light particles, because this would change the refractive index, which depends on the ratio of the speeds of light outside and inside the medium. Thus, light particles – or light rays – with different velocities would lead to a variable refractive index in the lenses of a telescope, at least according to the theory of refraction, but the refraction phenomenon is always the same. However, after FRESNEL introduced his theory of longitudinal aether waves, he noted that the motion of a telescope relative to the incident light waves should still cause a change in refraction even if the speed of the waves were constant. The problem was the following: since the telescope, when orbiting the sun with the Earth, sometimes moves laterally to the incident light beam of the star, the wave fronts hitting the Earth as plane waves can no longer be incident vertically from above, but would have to reach the lens at an oblique angle. According to the HUYGENS-FRESNEL principle, this should actually lead to additional refraction effects in *moving* glass lenses and translucent bodies – but these were apparently not detectable. So, before the wave fronts arrive, the telescope has already moved somewhere else, which makes a vertical incidence of the aether waves possible only in the rare cases when the Earth moves exactly towards or away from the star. FRESNEL therefore had to assume that longitudinal pulse waves must also be treated with a transversal component of motion to compensate for this additional effect. This is sometimes illustrated by the picture that the telescope must assume a certain lead angle in order for the wave fronts to enter the telescope perpendicularly. However, this idea makes no sense if we recall that BRADLEY used a zenith telescope that always pointed vertically upward. Thus FRESNEL was forced to introduce an additional assumption that ensured the invariance of the refraction effect even in moving systems. This was the so-called *aether entrainment hypothesis* in its original form, which was later also called *aether drag* (in reference to the air drag of moving bodies). This hypothesis states that a very small portion of the aether gas is entrained by the moving translucent matter (by glass prisms, lenses, water etc., but not by air), which should result in a slightly increased aether density that restores the normal refractive index.

So the problem might have two possible causes: on the one hand, the theory of refraction might be based on wrong assumptions, on the other hand, the theory of longitudinal waves might be wrong. Since, as we will see in a moment, the transverse wave theory has the same problem, it must be assumptions related to the mode of propagation of light itself. For example, there is the assumption of a constant speed of light in the aether. It is demanded on the one hand by the theory of refraction, but also by the longitudinal pulse wave theory of light. Because if it is about pulse waves, their propagation speed can no longer be influenced by the motion of the light source. As in the case of explosion and sound waves, it may even be that the source no longer exists after the emission of these waves. And as in the theory of sound, pressure and pulse waves, the movement of a receiver relative to the wave fronts would then have to change the time arrival rate of the wave fronts, called wave cycle or wave period. In the case of sound waves, this would be interpreted as a change in the locally effective speed and frequency (tone pitch) of the sound waves. And in the case of light waves, this would be interpreted as a change in the locally effective speed and frequency (color) of the light waves. So, theoretically, the motion of a light receiver should cause a changed refraction in glass bodies, which obviously was not the case. So while we can be fairly sure that the sound wave theory applies to reality (even if a new kinetic theory of gases were to incorporate pressure-dependent changes of the molecular structure), this is not quite so certain for the light wave theory, at least as far as the mode of propagation – the nature of that movement – is concerned.

FRESNEL's aether entrainment hypothesis also stated indirectly that normal matter, a telescope filled with air and the Earth itself move through a gaseous medium without interacting with it. So, on the one hand, the aether gas is supposed to fill a space, to be at rest and not to be disturbed by moving bodies. The matter does not act on the aether, the aether does not act on the matter. On the other hand, there is supposed to be an exception for this: only moving translucent matter (dielectrics, as we now know) are supposed to interact with the aether waves (light) and partly entrain it. This was a contradiction deeply embedded in the foundations of optical theory.

At first, FRESNEL, like YOUNG, had assumed longitudinal waves and could thus have established a unity of light and gaseous matter, if he had succeeded in solving two problems – the invariance of the refraction effect and the question of friction. Then the polarization effect came in between, which seemed to be explainable only with transverse waves. But the transverse wave concept was obviously no longer compatible with the original assumption that light waves are excitations of an extremely rarefied gas: if transverse waves can act with a very high speed over arbitrary large distances without losses because of the high elasticity of the medium, the aether must be virtually a solid. With the unclear constitution of the aether substance also the question about friction became superfluous: no planetary system could survive in such a material. So there had to be something wrong with either the idea of transverse waves (the explanation of polarization) or the propagation of light – or both.

At this point we can now think different. We have two, even three clues: The holistic divisibility of the integral molecule, the holistic divisibility of light and matter in the double-slit experiment, and the instantaneous remote actions of quantum theory, which are incompatible with the spirit of special relativity. So if we assume that light does not move at all in the sense of mechanics, but needs time for its structure formation, which is directly proportional to the distance between the light source and the receiver, we could well have to do with structure-forming processes of an ordinary gas with extremely low density. In agreement with AVOGADRO's division hypothesis, these can be interpreted as scale-free molecular cell division processes, but also as the formation of standing waves (or cells), which can have both local and global (aka non-local) character. So let us see if the problems of wave theory and quantum physics become more comprehensible from this perspective.

The initial problem was that the assumption of a constant speed of the incident light rays seemed to be indispensable for the explanation of refraction and stellar aberration, but led to contradictions in both wave theories as soon as a light-refracting body moves relative to these light rays. Since NEWTON, the theory of refraction has been based on the premise that a beam of white sunlight consists from the beginning of different colored components that are only spatially separated, split or dispersed by a glass prism. In wave theory, this assumption is maintained, but the components are now rays of light consisting not of particles but of waves whose color is defined by the frequency of these waves. Although it is known that the spectrum of rainbow colors, which is produced by refraction at a glass prism, is continuous, the idea is still held that the incident light beam consists of many – actually infinitely many – individual partial waves with finely graded colors, that is, frequencies. From this premise it follows that the color or frequency of the partial rays must be preserved when passing through a glass prism and only the wave period (which is a time unit) can change, which was interpreted as a change of the speed of light in the medium. The extent of refraction had to depend on the frequency of the light, since blue light is refracted more strongly than red light. In addition it depended on the physical properties of the material through which the light passed. The speed of light in translucent media such as glass, crystals or water becomes smaller than in aether, vacuum or air, because the effective wavelengths (aka wave cycles) become shorter as the density of the material becomes greater. The ratio of the speeds at the transition from one medium to another is expressed by the refractive index. It explains why a rod immersed in the water appears to have a kink. The kink in water is different from the kink in air, where it is practically zero.

In the case of stargazing, however, this no longer seemed to apply. As one was astonished to discover in 1872, the aberration constant remained the same when the telescope was filled with water (instead of air). The refraction did not change. This is exactly what FRESNEL had already predicted in 1818 and justified with his aether entrainment coefficient. However, the physical meaning of this refraction correction factor, which was supposed to depend on the velocity of the light refractive body, was still a mystery: How can a movement of the telescope around the sun cancel or reverse the different refraction of light in air and water, which undoubtedly takes place in the telescope? Surely this makes no sense at all, if this simply describes a change of reference frame, while the physical processes must remain the same. The problem immediately reminds of the FARADAY induction, which likewise can depend only on a relative motion, i.e. on a change of distance between the two bodies. Only then it becomes irrelevant from which perspective this physical process is described. However, electromagnetic induction had not yet been discovered at FRESNEL's time, and with it the fact that light is of electromagnetic nature and polarization a purely magnetic phenomenon. FRESNEL had obviously already recognized that this problem requires a symmetry of motion, as it was known at that time only from mechanics, where it was called *relative motion*. However, the task of transferring the principle of relative motion to dynamic processes (aka structural movements) of light seemed unsolvable as long as the nature of polarization in light and matter had not been elucidated.

After the introduction of the transverse wave model between 1821 and 1823, which was intended to explain the polarization phenomenon, FRESNEL had to adapt his entrainment hypothesis to the new model. He designed a sophisticated mathematical model that states that some transverse waves of the incident starlight are exactly compensated by the transverse component of the Earth's motion around the sun. In this way, he relates the refractive index to the velocity of the refractive body relative to the incident transverse waves. This should apply to water, glass lenses, glass prisms, and other translucent substances, with the exception of air and aether. So FRESNEL still assumed that relative motion to the (excited) aether gas changes the speed of light in moving translucent matter, which implied some kind of speed-dependent "friction" between light and bodies at the boundary layer or inside the body. If the relative velocity is zero, there is no "friction". This means that some transverse waves must interact with water or glass in such a way that refracting matter literally traps and entrains these light waves by adhesion or unknown interactions of a similar nature (which could well be branching processes). If this idea was correct, the relative speed of light in light-refracting media had to depend on whether the medium or body was moving in the direction of the light ray or in the opposite direction. The FRESNEL coefficient is then to be added to the velocity of light if the light wave propagates with the moving translucent medium and subtracted if it propagates against the moving translucent medium.

This immediately reminds us of the relative speed between ordinary moving bodies and gases, for example air. If the body moves against the stream of air, the relative speed between the two is higher. If it moves with the stream, correspondingly lower (if stream and body velocity are equal, the relative speed between the two is zero). So the FRESNEL coefficient, which hides an unknown interaction between translucent matter and light, now represents a correction factor for the refractive index, which depends on the speed of the light-refracting body or medium. But how is *this* speed determined at all?

Certainly not relative to an immobile aether gas or to an aquarium in which the gas rests, as one might now assume (and did assume at the time). Because then there is no such 'friction', which fits quite well with the conception that 'normal' bodies do not interact with the aether gas – as long as they do not emit or receive light. Whether material bodies are slowed down by friction in an extremely rarefied gas, which even from a modern point of view resembles a *high vacuum*, is thus a completely different question. The answer depends on how we interpret friction at the molecular level. So this speed (of the material body) can be defined only relative to those movements of the aether gas, which we call excitations, light or polarized states. This fact seems to have been overlooked more than once. Until today we try to understand these movements as unidirectional motion of light beams and waves, but encounter mysterious difficulties which seem to overwhelm our logic. So FRESNEL's entrainment coefficient was mainly a speed correction factor that determines the speed of light in translucent bodies as a function on its own speed *relative to the 'free' speed of light*.

In 1851, FIZEAU performed a speed-of-light interference experiment with two water-filled tubes to prove FRESNEL's theory. He branched a light beam with a beam splitter into two partial rays: one moves in one tube with the water flow, the other in the other tube against the water flow. After that, the rays are reunited so that they can superimpose. As predicted by FRESNEL, the interference pattern shifts as the water flows, indicating a mirror-symmetric change in the wave periods of the partial rays. According to wave theory, this was interpreted as a difference in the speed of light. The speed deviation of the partial rays was equal in amount: what the one ray had more in speed, the other one had less. Something similar, or possibly even the same, we already know from the magnetic branched light rays in FARADAY's rotation experiment and ZEEMAN's spectral experiments, where the wave periods or wave cycles of the branched partial rays also change mirror-symmetrically. The speed deviation corresponded exactly to FRESNEL's aether entrainment coefficient, which thus actually proved to be a correction factor for the refraction of light waves (and light rays) in moving translucent media. This seemed to prove conclusively that moving translucent liquids and solids carry with them a small portion of the aether – more precisely, a tiny fraction of the excited aether gas we call 'light' even if we can't see it. In other words, FRESNEL's speed correction factor seems to describe a physical interaction of light with matter which, like FARADAY induction, can only be understood as relative motion or structural movement between light and matter. In this context, branching processes have apparently never been thought of.

In all these cases, we are dealing with interactions between light and matter in which the normally constant wavelengths of the emitted light change. According to the theory, the changes consist of shortening or stretching of the wave cycles, which changes the effective wavelength. Since MALUS and FRESNEL we know that the property of polarization becomes noticeable only after reflection and refraction processes, which obviously always have to do with branching processes of light rays, which, however, have always been considered purely phenomenologically so far. Since MAXWELL, we know that a wavelength is not only a unit of length containing two opposite polarization states, but also a unit of time describing the period in which the aether gas – or light in the abstract – has passed through two opposite polarization states. And since ZEEMAN we know that magnets cause branching processes that produce opposite polarization states of the partial rays and mirror-symmetric changes in their wave cycles. At that time, the changes in matter itself were still out of consideration – ZEEMAN's experiments had just opened the door to this unknown world.

So the duration of the wave cycles is not always the same – sometimes the interaction with matter runs faster, sometimes slower. The question is whether we must necessarily associate these wavelengths with a *velocity* of light, i.e. with unidirectional motion and change of location. Astonishing is that even the motion of the bodies should have an influence on the interaction cycles with light. Even stranger, sometimes there is a difference in speed – a change of wave period – due to relative motion, sometimes there is not.

In ordinary gases, the relative speed between the longitudinal pressure waves and the body changes the effective wavelengths or wave cycles of the wave fronts hitting the body. This is described by the Doppler effect. In the case of stellar aberration the changes of wave period are explained by a *motion* of the Earth relative to the *motion* of transversal wave fronts hitting the body. In the case of refraction in moving translucent bodies and in FIZEAU's water experiment they are explained by the *motion* of light waves relative to the *motion* of a light-refracting medium. But in the MICHELSON-MORLEY experiment, there is no change of the wave periods detectable and thus no speed difference – neither between the branched partial waves nor between them and the surrounding aether medium. Thus, there is no *motion* of the light waves relative to the *motion* of the body or a medium. In the case of diffraction (bending, branching and interference), the speed of light is supposed to remain constant, but the frequency of the branched partial waves changes, which means the same: a change in the effective wavelength or wave period. And in the experiments of ZEEMAN we observe a magnetic induced branching of the partial rays which differ mirror-symmetric by the same amount of wavelength or frequency, but show no speed difference.

All this led to considerable confusion. Was the aether now a gaseous substance, which did not take part in the movement of the bodies? Or was it partly dragged along? The contradictions began to pile up.

To explain the null-result of the MICHELSON-MORLEY experiment without FRESNEL's aether entrainment conception but in agreement with FIZEAU's water interference experiment, A. LORENTZ attempted to derive the lightspeed correction factor in a different way. In 1897, after years of pondering, he realized that the problem of transverse wave theory in explaining stellar aberration and the MICHELSON-MORLEY experiments was to derive the invariance of the laws of refraction for systems at rest and in motion from a symmetry of the equations of motion. His most ingenious insight was that this requires not only a symmetric, but a *mirror symmetric* picture. The description of the process, in which two material bodies or spatial reference systems interact by light, must consist of two states, which 'correspond' to each other, thus are equivalent, but can be treated only together with a mirror-symmetrical coordinate system (at this point already hides the hint to the holistic character of the interaction and enantiomorphic field states). The transition from one spatial reference system to the other is represented by a mathematical vector operation that describes a *spatial* mirroring of the coordinate system. Since LORENTZ referred to the HERTZ-HEAVISIDE version of the MAXWELL equations (which excluded the magnetic polarization of the aether gas), it was a vector with only four components instead of six. This spatial mirroring could be expressed quite simply with elementary algebra, which – physically interpreted – models a topological space-like inversion (of a 'four-vector space') that transforms a right-handed reference frame into a left-handed one. Exactly this can be illustrated with the picture in which a glove is turned inside out. This is the nature of the mathematical operation which POINCARÉ christened *Lorentz transformation* in 1904.

In other words, the two material bodies and spatial reference frames should actually be *ontological* mirror images of each other and form an enantiomorphic system. But LORENTZ could not see that far yet. For him these were mainly mathematical tools whose physical sense was not at all clear yet. In this spatial inversion transformation then FRESNEL's correction factor for the local lightspeed comes into play. It was difficult, however, to justify this correction factor physically: if one system moved in the direction of the light beam, the speed of the moving system had to be subtracted from the speed of light to be able to determine the local speed of light in the moving system. Mirror-symmetrically, the other system then moved relative to the first one in the opposite direction, i.e. against the light beam, therefore one had to add the speed of the moving system to the speed of light. The FRESNEL correction factor thus describes mirror-symmetrical deviations from the constant speed of light. The deviations depend on the relative speed between the two systems, which in turn is defined relative to the speed of the 'free' light. This looks like nonlinear feedback effects that must have to do with the structure of light and matter.

Although in such a symmetrical or relativistic picture actually neither of the two systems can be preferred, as GALILEO had already made clear two hundred fifty years earlier with his principle of relative motion, LORENTZ considered one of the two bodies or reference frames as the absolute, immobile space, i.e. as an imagined aquarium, filled with a gas of unknown nature, and the other as the true moving body and reference frame. This was a logical, quite natural concept, which had already proven itself thousands of times in the treatment of ordinary gases. It would be desirable (and would simplify many things) if we could apply this conception to the nature of light as well. Consequently, it should be theoretically possible to prove experimentally the motion of bodies through the aether with the Doppler effect. If one follows the principle of relative motion, one can conversely also consider the body as immobile and should then be able to prove an aether wind. However, this would only be true if the light consisted of longitudinal waves. The motion of the body through the gas, or the motion of the gas relative to the body, would refer in this picture to a gas which does not interact structurally with the body. This is the only reason why we can treat the problem purely kinematicaly as a relative motion. The subtleties are hidden behind the term friction, which practically assumes the role of a correction factor summarizing the microscopic processes in the boundary layer.

In the case of the light, the situation seems to be similar, although light is supposed to consist of transversal waves, which nevertheless are supposed to propagate spherically and rectilinearly like longitudinal waves in all space directions. Now one would like to prove a relative motion of the matter to the light waves or the light waves to the matter, which would have to result in a change of the effective wavelength in the matter. That is, the matter must interact structurally with light. In analogy to the effect of sound waves on moving bodies, the interaction was initially interpreted as a change in the incoming effective wave periods, which then stimulates the receiver membrane to vibrate synchronously. In the case of light, however, the receiver membrane is not passive at all, as refraction and spectral experiments show: refraction shows that the interaction of light with matter makes polarizing branching processes of light that only seem to have nothing to do with structural changes in matter. And the splitting of spectral lines shows that matter itself does not passively and synchronously resonate, but actively reacts with structural changes of the molecular structure. This is the essential difference between these two cases, apart from the wave models.

The MICHELSON experiment had shown that a light beam branches into two partial rays by refraction at a glass plate, whereby the wavelengths and frequencies of the two partial rays did not differ. Consequently, the motion of the Earth could have no influence on the velocity of light of the partial rays, although at least the partial ray in the direction of motion should have indicated a change of the wave period and thus a change of its speed. Obviously, FRESNEL's lightspeed correction factor had to act in this partial ray to compensate the change which occurs by the relative motion to the aether gas. Did one have to consider a change in the structure of matter that had not been considered before? Could MAXWELL's theory be modified by the FRESNEL factor in such a way that the motion of the interferometer, which must lead to a structural change in ordinary matter, is compensated by certain physical processes in such a way that the matter appears unaffected despite real structural changes?

This was the point at which ANTOON LORENTZ integrated the ideas of FITZGERALD and LARMOR into his concept of inversion transformations. By 1904, LORENTZ had completely revised his mathematical models and designed an extended MAXWELL-HERTZ-LORENTZ theory that explained the null result of the MICHELSON-MORLEY experiment with dynamical processes that represented mirror-symmetrical scale changes in the structure of light and matter. To this end, he postulated an *electrodynamic* force that allowed FRESNEL's aether entrainment hypothesis to be ignored and interpreted as a kinematic speed correction factor describing structural changes in matter as changes in the orbital periods of electrons relative to incident and emitted light waves. However, this was not entirely free of contradictions: Recall that AMPERE's notion of an electrodynamic force was based on the premise that there are instantaneous actions over arbitrarily large distances as in the theory of NEWTON. This is, of course, incompatible with a field theory based on local near-actions, the corresponding wave model, and a constant speed of light. This electrodynamic – aka polar magnetic – force should act exclusively on the interferometer arm pointing in the direction of motion and should cause a *length contraction* and *stretching of the 'local time'*.

Of course, these assumptions pursued a very specific purpose: If length and time scales change inversely proportional to each other, the quotient of length and time units, which defines the speed, always remains constant – and thus also the speed of light. In this way LORENTZ integrated the ideas of FITZGERALD and LARMOR and could ensure, at least theoretically, that the branched partial rays of the moving interferometer each propagate with the velocity c, which in turn seems to confirm the validity of the MAXWELL-HERTZ equations. The price for this was a weird shortening of the interferometer arm in the direction of motion *in reality*, which was supposed to be caused by magnetic forces generated by the motion alone. With this construct, LORENTZ arrived in 1904 at an extended electromagnetic theory, in which the speed of light always appears invariant for moving bodies in all spatial directions, exactly as in the reference frame of a non-moving aether gas volume.

Closing words to the first volume

Obviously, natural science does not develop linearly, but approaches truth in iterative loops. Around 1900, the separation of Nature into matter and (aether) fields still seemed to be a promising approach. But understanding the interaction – and unity – of matter and light by combining electromagnetic theory with mechanics and the atom hypothesis proved to be difficult. Since LORENTZ, theoretical physics has had to grapple with two apparently different entities: with matter, a material substance that has mass and is treated as a body in the sense of mechanics, and with electromagnetic fields, which should not have mass and were therefore considered 'non-material'. The electromagnetic field mediates effects via structural changes of the aether gas, which basically had to be (not yet very well understood) electric and magnetic polarization processes. Thus, there was no longer any reason to assume remote, instantaneous effects as in the theory of gravity and electrodynamics. So for many physicists, everything seemed to be fine with physics around 1900 – after all, much had been achieved in the previous century. The aether wave theory of light was successfully united with electromagnetism, the existence of the aether was almost proven; even the mystery of the atom seemed within reach. Some physicists even believed that there was hardly anything new to discover: one simply had to treat matter with Newtonian mechanics, light with MAXWELL's electromagnetic theory and moving electric particles with the MAXWELL-LORENTZ equations. And yet there were still a lot of unsolved problems: the concept of motion in mechanics and the definition of the speed of light encountered inexplicable contradictions in the explanation of the MICHELSON-MORLEY experiment. The nature of the aether substance and the meaning of FRESNEL's aether entrainment coefficient were still unclear, as was the question of why MAXWELL's model of FARADAY induction could not really reflect the principle of relative motion. HERTZ and LORENTZ had taken over the assumption of a unidirectional motion of light – and electrons – directly from MAXWELL without questioning it. EINSTEIN took over the definition of the speed of light from MAXWELL, but only after years of pondering in which all other ideas failed. In this way, the symmetry or relativity of motion, which was already built into WEBER's constant c, was practically lost – which at first nobody noticed, until EINSTEIN came along.

In the first volume of "Solving the Quantum Puzzle. Understanding the Cognitive Problem", we became familiar with the history of ideas in 19th century physics, learned about the nature philosophical roots of the quantum problem, and got a feel for how new physical conceptions enter the world. Obviously, the process of cognition is by no means linear, but is an interplay between groping search and creative design, in which conceptions, models, hypotheses and constructive theories must be repeatedly tested against reality and experimental facts. The open questions about the nature of polarization and the true nature of light and matter led, among other things, to relativity and quantum theory, which shaped the physical concepts of the 20th century, especially mathematically. However, the actual cognitive problems of physics could not be clarified until today. For hitherto inexplicable reasons, there seemed to exist a *sonic barrier of cognition* in physics that seemed insurmountable to ordinary mortals. Yet now we know what this barrier consisted of: the atomic paradigm and the inability to recognize and physically acknowledge the failure of this paradigm.

In the second volume, we will continue to explore these questions. We will follow the development of special relativity and quantum physics, analyze the key experiments, and become better acquainted with BOHR's Copenhagen interpretation and EINSTEIN's views. We will witness how the body idea of our mind (and of mechanics), the atom hypothesis, the wave theory, and the definition of the speed of light fail in these experiments, finally leading theoretical physicists to throw in the towel in 1927 in utter despair. Since then, it is considered impossible on principle to solve the quantum puzzle, i.e., wave-particle paradox, and to design a new, realistic understanding of nature in the original sense of scientific enlightenment.

Don't believe in it. The second volume will convince you on the basis of clear experimental evidence – which no physicist doubts – that it is not only possible but urgently necessary to design a new physical world view that does without the atom hypothesis. I think that the second volume can be published in three to six months. I hope you will stay tuned!

Mario Wingert, October 2023

List of Figures

Literature & Sources

Autorenkollektiv: **Struktur der Materie.** VEB Bibliographisches Institut Leipzig 1982

Avogadro, Amedeo: **Essay on a Manner of Determining the Relative Masses of the Elementary Molecules of Bodies, and the Proportions in Which They Enter into These Compounds.** 1811 web.lemoyne.edu/~giunta/ea/AVOGADROann.HTML
Deutsche Übersetzung in: Wilhelm Ostwald: **Die Grundlagen der Atomtheorie.** Ostwalds Klassiker 3/8; Leipzig 1902

Assis & Chaib: **Ampere's Electrodynamics.** C. Roy Keys Inc., Montreal 2015

Baggott, Jim: **Farewell To Reality. How Modern Physics Has Betrayed the Search for Scientific Truth.** Pegasus Books NY, 2014

Born, Max: **The Statistical Interpretation of Quantum Mechanics.** Nobel Lecture, 1954

Baeyer, Hans-Christian von: **Das Atom in der Falle.** Rowohlt Taschenbuch Verlag 1996 (1992)

Cramer, J.: **The Transactional Interpretation of Quantum Mechanics** (www.npl.washington.edu/ti)

Davies, Paul & Brown, Julian: **The Ghost in the Atom.** Cambridge University Press 1986

Einstein, Albert: **Über die spezielle und allgemeine Relativitätstheorie**, Vieweg Verlag 1979 (1916)

Einstein, Albert & Infeld, Leopold: **Die Evolution der Physik.** Rowohlt Taschenbuch 1998 (1938)

Feynman, Richard: **Vom Wesen physikalischer Gesetze.** Piper Verlag 2001 (1965)

Feynman, Richard: **QED. Die seltsame Theorie des Lichts und der Materie.** Piper Verlag 2002 (1965)

Feynman, Richard: **Physikalische Fingerübungen.** Piper Verlag 2006 (1963)

Fölsing, Albrecht: **Albert Einstein. Eine Biographie.** Suhrkamp Taschenbuch 1995

Gleick, James: **Chaos. Die Ordnung des Universums.** Droemersche Verlagsanstalt 1990 (1987)

Gribbin, John: **Auf der Suche nach Schrödingers Katze.** Piper Verlag 2000 (1985)

Gribbin, John: **Schrödingers Kätzchen.** Fischer Taschenbuch Verlag 1999 (1995)

Heisenberg, Werner: **Physik und Philosophie.** Ullstein Verlag 1968

Heisenberg, Werner: **Der Teil und das Ganze;** Piper 2002 (1969)

Hertz, Heinrich: **Untersuchungen über die Ausbreitung der elektrischen Kraft.** Leipzig 1891

Hertz, Heinrich: **Electric Waves.** London 1893

Hey, Tony & Walters, Patrick: **Das Quantenuniversum.** Spektrum Akademischer Verlag 1998

Hund, Friedrich: **Zur Deutung der Molekülspektren** 1/2, ZF f. Physik 1927

Karlsch, Rainer: **Hitlers Bombe**, Deutsche Verlagsanstalt 2005

Kumar, Manjit: **Quantum. Einstein, Bohr and the great Debate about the Nature of Reality**, 2010

Maxwell, James Clerk: **A Dynamical Theory of the Electromagnetic Field.** Phil. Transactions, 1864

Meadows, Donella: **Leverage Points. Places to Intervene.** 1997

Mulliken, Robert S.: **Electronic Structures of Polyatomic Molecules and Valence** (1). Phys. Rev. 1932

Mulliken, Robert S.: **Spectroscopy, molecular orbitals, and chemical bonding.** (Nobel Lect. 1966)

Pais, Abraham: **Raffiniert ist der Herrgott.** A. Einstein. Eine wissenschaftl. Biographie. Spektrum 2000

Penrose, Roger: **Computerdenken.** Spektrum Akademischer Verlag 1991 (The Emperor's New Mind)

Penrose, Roger: **Schatten des Geistes.** Spektrum Akademischer Verlag 1995 (1994)

Rae, Alastair: **Quantenphysik: Illusion oder Realität?** Philipp Reclam jun., Stuttgart 1996 (1986)

Rosenblum & Kuttner: **Quantum Enigma. Physics Encounters Consciousness**, Duckworth 2011

Schrödinger, Erwin: **Quantisierung als Eigenwertproblem. Vierte Mitteilung.** Juni 1926

Smolin, Lee: **Trouble with Physics**, Houghton Mifflin 2009

Stachel, John: **Einsteins Annus mirabilis.** Rowohlt Taschenbuch 2001 (1998)

Wingert, Mario: **Einsteins Vermächtnis: Die Revolution der Physik.** Die Auflösung des Welle/Teilchen-Paradoxons. 2003 (Book on Demand, Norderstedt, Germany)

Wingert, Mario: **Quantum Top Secret - Die Lösung des Quantenrätsels.** 2008 (Book on Demand)
Wick, David: **The Infamous Boundary.** Copernicus, New York 1996
Zeeman, Peter: **On the Influence of Magnetism on the Nature of the Light Emitted by a Substance,** 1897
Zeilinger, Anton: **Einsteins Schleier. Die neue Welt der Quantenphysik.** Verlag C.H.Beck 2003
Zeh, H. Dieter: **Physik ohne Realität: Tiefsinn oder Wahnsinn?** Springer Verlag 2012